A HYPER CONNECTED WORLD

802.11ax
AND
THE NEXT
GENERATION
WiFi

Pablo Aguilera, PhD

First Printing, 2018

KDP ISBN 9781981002726

pablo.aguilera@galgus.net

Galgus (AOIFE Solutions S.L.)

+34 955 382 328

Calle Itálica 1, 1º. 41900 Camas, Seville. Spain.

www.galgus.net

This book is dedicated to someone who has not yet arrived.

To the evidence that growing up isn't that bad.

To Elena.

Contents

3 - An overview of 802.11ax.................... 53

4 - 802.11ax: PHY layer features 73

5 - 802.11ax: MAC layer features 89

6 - Radio Resource Management..................... 109

1
Background

1.1 - Introduction

The IEEE (Institute of Electrical and Electronics Engineers) 802.11 family protocols (commercially known as WiFi) is one of the most influential technologies in today's society. It was designed in the last years of the XX Century to substitute Ethernet infrastructure in offices. Two decades later, this technology has been successfully deployed all over the world, providing an affordable solution in vertical markets like hotels, aircraft, ships, hospitals, stadiums, or shopping centers. WiFi networks enjoy its best health ever thanks to its use of the shared non-licensed electromagnetic spectrum around 2.4 and 5 GHz frequency bands. However, **WiFi enters as a mature technology in its third decade,** facing important challenges as a result of current telecommunication trends and the needs and demands of the XXI Century.

The proliferation of WiFi networks in almost every place brings to the scene the law of diminishing returns, where a larger WiFi infrastructure does not imply a better *QoS (Quality of Service).* As we will see later, the shared media access in WiFi is asynchronous, so it is usual that two transmitted frames collide in the receiver, making impossible the decoding and forcing a retransmission. **The solution consists, again, on a better use of radio and logic resources** (transmitted power, time slots, frequency channels, etc.), in order to maximize *SNIR (Signal to Noise and Interference Ratio).* This is aligned with some of the modern social concerns, such as higher energetic efficiency, lesser electromagnetic radiation, and lesser power consumption.

As a result, it is evident that the next 802.11 release must address important issues to solve the aforementioned challenges. For that reason, in the last years, the IEEE *HEW (High-Efficiency WLAN)* Study Group has been passionately working on new and innovative technologies for the wireless communications of tomorrow, while saving the backward compatibility with current WiFi devices. As we will see in the following Chapters, the proposed solutions suppose a breakthrough in the well-established paradigm of WiFi communications.

1.2 - Evolution

Wireless networks based upon IEEE 802.11 standard, have experienced an overwhelming success since its first versions, two decades ago. In practice, the data rates have been multiplied by 1000, providing smart, reliable, and fast RANs. Table 1 summarizes the main features of the main 802.11 releases.

	Year	Throughput (max)	Frecuency band	Order	Modulation	Bandwidth	MIMO
802.11a	1999	54 Mbps	5 Ghz	64-QAM	OFDM	20 Mhz	1x1
802.11b	1999	11 Mbps	2.4 Ghz	-	DSSS	20 Mhz	1x1
802.11g	2003	54 Mbps	2.4 Ghz	64-QAM	OFDM	20 Mhz	1x1
802.11n	2009	65-450 Mbps	2.4/5 Ghz	64-QAM	OFDM	20, 40 Mhz	up to 3x3
802.11ac	2013	290-3600 Mbps	5 Ghz	256-QAM	OFDM	20, 40, 80,160 Mhz	up to 4x4 downlink MU
802.11ax	2018	600-8000 Mbps	2.4/5 Ghz	1024-QAM	OFDMA	20, 40, 80,160 Mhz	Up to 8x8 downlink/uplink MU

Table 1. Comparison between 802.11 wireless standards.

Despite early intents, **802.11a** was the first successful standard published in 1999. The proposed modulation was *OFDM (Orthogonal Frequency Division Multiplexing)*, with 52 subcarriers for the *PHY (Physical layer)*. It operated on 5 GHz frequency band, and it was stated to achieve up to 54 Mbps. Nevertheless, its implementation was an enormous technological challenge, so the release was delayed several months, and gave time to a strong establishment of 802.11b.

On his part, **802.11b** appeared the same year. It worked on the 2.4 GHz band but using spread spectrum techniques like *DSSS (Direct-Sequence Spread Spectrum)* to access the radio channel. It achieved a rate up to 11 Mbps and had a larger coverage area than 802.11a, but unfortunately, it was subject to more interference. The reason was the shared spectrum with cordless phones, Bluetooth, microwave ovens, and many other technologies that operate within the same frequency band.

802.11g arrived in 2003, merging the best characteristics from both standards: it operated on the 2.4 GHz band but used OFDM modulation to access the radio channel, obtaining again up to 54 Mbps. It was backward compatible with previous technologies and promoted the settlement of WLAN technologies as a reliable and cheap substitute for Ethernet. Dual band a, b, and g products spread through the market and accelerated in a notable manner the penetration of WiFi devices within the 802.11 framework.

Being WiFi the absolute king of local wireless communications, **802.11n** was released in 2009. Also known as *HT (High Throughput)*, it announced data rates up to 600 Mbps. To achieve this, it proposed for the first time the use of multiple transmission and reception antennas, which was known as *MIMO (Multiple-Input, Multiple-Output)*. It also allowed to aggregate legacy 20 MHz channels in pairs (40 MHz) under user request. This channel bonding policy was going to become a frequent improvement in the next releases of the standard. In addition, it also proposed space-time coding and beamforming (which was not widely implemented due to technical difficulties) to improve robustness and coverage area of the radio signal. Frame aggregation helped to reduce overhead by grouping data under the same header. So much expectation was created that 802.11n compatible products arrived at the market even before the standardization process was completed.

802.11ac is the latest version of the WiFi standard for WLANs. It was approved in 2013, and it defines how the standard implements *VHT (Very High Throughput)* networks, with the aim of achieving data rates of several Gigabits per second. It operates exclusively on the 5 GHz band, and it allows bonded channels of 80 and even 160 MHz. In addition, it raises the modulation order in optimal conditions up to 256-*QAM (Quadrature Amplitude Modulation)*. However, its biggest step forward is the extended use of MIMO, supporting up to 4 antennas transmitting and receiving, adaptive beamforming, as well as downlink *MU-MIMO (MultiUser MIMO)*. Later in this book, we will review these technologies. Some of these improvements were a major engineering challenge, so the market release has been split into two phases: **Wave 1** at the end of 2013 with the basic features (which are a substantial improvement over 802.11n), and **Wave 2** with advanced features in 2016.

Finally, in early 2018, there is a stable Draft for **802.11ax**, the "next big thing" in WiFi[1]. The first chips have been built and tested before the official release of the new amendment, but the current Draft main lines shall remain unaltered[2]. It is important to remark that 802.11ax devices will reach the market[3] in different *Waves*, as happened with 802.11ac. During 2018, we will see the first commercial products deployed and working in the real world.

This book aims to serve as a deep review of the main features of this promising release, which brings **the most important changes seen for WiFi in decades**. The author hopes that it will serve engineers, scientists, technicians, network architects, and managers to achieve a full understanding of the amazing improvement of the new WiFi standard.

---

[1] *https://trends.google.com/trends/explore?q=802.11ax*

[2] *http://www.ieee802.org/11/Reports/tgax_update.htm*

[3] *https://www.cavium.com/Quantenna-Announces-World's-First-802-11ax-Wi-Fi-Solution.html*

1.3 - Data Rates

In the late 90s, the wireless network designs were based upon **coverage** constraints. In the future, the focus will be on **energy** consumption and low latency. However, during the last decade, the engineers, researchers, and network architects are designing WiFi networks to maximize the expected **throughput** of a given wireless infrastructure. This may be seen in Figure 1, where the throughput for one user is shown.

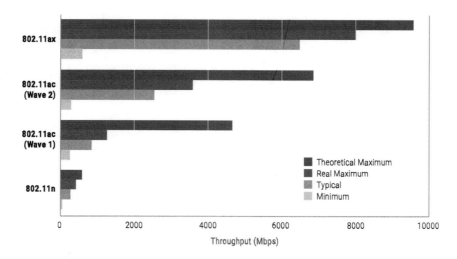

Figure 1. Expected throughput on different 802.11 standards.

As expected, the throughput rises substantially when introducing better modulations, larger channel bandwidth, and more transmitter-receiver antennas. On that, it shall be noted that this effective throughput is calculated for the better channel conditions, and for only one user. However, we can get closer to this upper bound when using the newest devices available.

To finish this Chapter, we will describe a big picture of the last features introduced in WiFi, specifically with respect to previous amendments of the 802.11 standards. This will give the reader an overview of the state of the art before introducing the newest 802.11ax features.

1.4 - State of the Art: 802.11ac Wave 2

1.4.1 - Main features with respect to 802.11ac Wave 1

Wave 2 is the second phase of 802.11ac devices, which has flood the market in 2017. Nowadays, devices that reach the consumers in this second phase of the standard, shall be prepared to implement the most impressive characteristics of 802.11ac[4]. Although some are mandatory, most of them will be optional. And that is the field where the vendors shall differentiate their proprietary solutions. We may summarize that, in 802.11ac Wave 2 devices:

a) 20, 40 MHz, and 80 MHz channels may be merged if they are contiguous in the spectrum (channel bonding) to create a **160 MHz channel**. This is only possible in the 5 GHz band, due to the inherent saturation of the lower 2.4 GHz band. As you may think, this is one of the reasons of 802.11ac for not operating on the legacy spectrum. With Wave 2, in addition, mechanisms for aggregating two non-adjacent channels of 80 MHz are included (the 80+80 mode).

b) By raising the modulation from 64-QAM (6 bits per symbol) to **256-QAM** (8 bits per symbol), one can bear 33% more information on the same carriers. This is only achievable if the channel conditions are optimal, from an *SNR (Signal to Noise Ratio)* point of view. A high SNR usually means that the user is close to the AP and without obstacles between them.

c) The use of multiple antennas shall bring several and diverse benefits. Before Wave 2, 802.11n and 802.11ac used MIMO for:

- Space-time coding for error correction.

---

[4] *Gast, Matthew S. "802.11 ac: A Survival Guide: Wi-Fi at Gigabit and Beyond". O'Reilly Media, Inc., 2013.*

- *SDM (Spatial Division Multiplexing)* for improving the throughput of one user.

- Beamforming for raising the coverage area on one concrete zone.

Channelization and modulation improvements are relatively easy to implement as an evolution of 802.11n and 802.11ac Wave 1, and the same applies to the basic MIMO operations aforementioned before.

However, Wave 2 devices that appear on the market have the chance to add downlink **MU-MIMO** (from the AP to the user terminals). This allows, for the first time in WiFi networks, to transmit to several users simultaneously, within the same collision domain, without interference. As an analogy of Ethernet wired networks, MU-MIMO[5] transforms the WiFi radio channel from one hub (where only one user may receive at a given time) to a switch, where there is a certain logic so users shall receive different data simultaneously. Note that if we have enough antennas, we may combine several of those paradigms to maximize the performance of our network.

1.4.2 - Wider channels: 160 MHz

As in previous amendments based on OFDM, 802.11ac Wave 2 divides the available channel onto sub-carriers, each one with 312.5 kHz of bandwidth. They bear an independent bit sequence. Most of them transport user data, but some are dedicated to pilot symbols (known both for the transmitter and the receiver), which help on channel quality measurements.

From 20 MHz legacy channels in 820.11a, the standard passed to 40 MHz in 802.11n, and later to 80 MHz in 802.11ac Wave 1. However, Wave 2 proposes to double the available bandwidth to build a bonded channel of 160 MHz. In

---

[5] *Ngo, Hien Quoc, Erik G. Larsson, and Thomas L. Marzetta. "Energy and spectral efficiency of very large multiuser MIMO systems." IEEE Transactions on Communications 61, no. 4 (2013): 1436-1449.*

addition, it presents an 80+80 MHz mode to aggregate non-adjacent channels, due to the lack of contiguous free spectrum in the 5 GHz band. Note that several types of **civil and military radars work on that band**, and their use has priority. Figure 2 shows the available channel in Europe and Japan on the 5 GHz band.

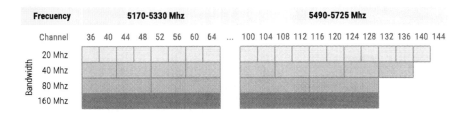

Figure 2. 802.11 channels on the 5 GHz band for Europe (and Japan).

The drawback of using 80+80 mode is that, as the channels are far from each other in the spectrum, the device must have duplicated transmission chains, raising considerably the cost and complexity of a deployment. Most of the vendors are sacrificing spatial diversity of MIMO, by using half of the RF chains for each one of the non-adjacent 80 MHz channels. An AP with 4 antennas shall run on MIMO 4x4 mode (four antennas for transmitting and receiving), unless when it needs the maximum bandwidth and requests 80+80 MHz. In that case, it shall run in MIMO 2x2 mode for each one of the channels.

1.4.3 - High-order modulations: 256-QAM

The increase of the quality of electronics in the RF chains, as well as the channel probing mechanisms, has favored the inclusion of high order modulations like 256-QAM. In QAM modulations, the bit word to be transmitted is mapped to one of the possible symbols of the alphabet. With 256 symbols, we may encode 8 bits within each transmitted waveform, instead of 6 bits (64-QAM modulation), 4 bits (16-QAM modulation), or 2 bits (QPSK modulation). This represents an improvement of at least 33% in spectral efficiency against Wave 1 and legacy releases.

Thanks to this, two new values of the *MCS (Modulation and Coding Scheme)* index are available. This physical layer index indicates the modulation and level of protection against errors (code rate) to be used in the transmission. This rate broadly specifies the portion of user information versus the total bits transmitted (the rest is filled with protection bits). The higher this rate (up to 1), the less conservative the transmission, and the better it is supposed to be the radio channel. The assignment has been significantly simplified with regard to 802.11n and can take values between 0 and 9. The last two values (8 and 9) are only unlocked with Wave 2 since they imply a 256-QAM modulation. Table 2 summarizes the MCS index values for 802.11ac.

MCS index	Wave 1 and 2								Only Wave 2	
	0	1	2	3	4	5	6	7	8	9
Modulation	BPSK	QPSK	QPSK	16-QAM	16-QAM	64-QAM	64-QAM	64-QAM	256-QAM	256-QAM
Bits/symbol	1	2	2	4	4	4	6	6	8	8
Code rate	1/2	1/2	3/4	1/2	3/4	2/3	3/4	5/6	3/4	5/6

Table 2. MCS index, modulation, and code rate in 802.11ac waves.

Why not go further and raise the modulation to greater orders? This is something that belongs to 802.11ax, so it will be discussed later in this book.

1.4.4 - Downlink Multiuser-MIMO

Most of the efforts that have been made over the years to implement Wave 2 solutions on 802.11ac devices are intended to equip APs with downlink MU-MIMO. Certainly, **multi-antenna and multi-user communications are capable of separating the collision domains within the radio channel,** steering signal where each one of the client terminals is located. To do this, more powerful processors are needed. They shall estimate the state of the channel and adapt the streams of the different transmitting chains in order to form a specific interference pattern.

So far, the MIMO paradigm used in 802.11ac Wave 1 standard was known as SU-MIMO (Single User MIMO), where the AP served data to an exclusive client by using all its antennas to multiplex several spatial streams to the receiver. There was forbidden to send different data to two or three different users at the same time. At most, **by using beamforming,** one may steer the energy lobe in the interesting direction, and thus gain a few decibels of SNR at the receiver front-end. This shall improve the MCS, and thus the throughput of a single user. However, with MU-MIMO, we can send data to all clients simultaneously, steering different bit streams on the directions of each user.

MU-MIMO unleashes the potential of multiple antenna devices, **by creating narrow and disjointed collision domains.** In advanced Chapters of this book, we will see how MU-MIMO may be also implemented in the uplink (from the STAs to the AP), thanks to the improvements of 802.11ax. Figure 3 compares the MIMO features of different 802.11 amendments.

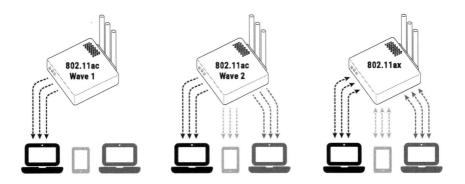

Figure 3. Comparison between SU-MIMO (left) and MU-MIMO (center and right) paradigms. With several antennas, one can serve several users at the same time, as long as they are sufficiently separated to form the proper spatial patterns.

1.5 - Conclusions

In this Chapter, we have analyzed the promising improvements introduced by the second phase of 802.11ac devices, namely Wave 2, in the context of advanced wireless communications. At this point, it should be clear that 802.11ac Wave 2 presents very attractive enhancements for a WiFi network in terms of throughput, spectral efficiency, and multiuser service. Note that some of Wave 2 most important features, such as MU-MIMO, require **terminals on the other end that understand and implement these new features**. Other features may be achieved by client software updates, or simply not be available for low-end devices.

As the reader may notice in the next Chapters, some of the features introduced by 802.11ax follow the same direction started with 802.11ac Wave 2. Other features open a new area of research and suppose a deeper breakthrough in what WiFi devices were doing until the release of this amendment. Of course, **all of these features have been designed to preserve backward compatibility**, so legacy devices will be able to connect to future WiFi networks, but without enjoying the latest advances of the standard.

In the next two Chapters, the author presents an overview of the main challenges that 802.11ax is facing. This will help the reader to visualize the needs and propose of this successful technology, and to understand why the HEW Study Group has introduced such new features in WiFi. Although some readers may be tempted to move directly to Chapter 3, we strongly recommend to review Chapter 2 before moving forward. It offers a different vision of the WiFi standard, which aims to clarify some key aspects that are usually overlooked in other technical books.

2

How WiFi works

2.1 - Introduction

In this Chapter, we will give an overview of the main characteristics of WiFi communications. The 802.11 standard defined a series of frames and operations for a successful wireless transmission between nodes. From the author's experience and from an educational point of view, it is better to introduce the pillars of the standard by showing how it was designed with two objectives in mind: simplicity and functionality. How it solved the particularities of asynchronous wireless communications. For that reason, **important frames and operations will be introduced sequentially along the following Sections. They will arise as they solve a specific issue in the standard**. Nevertheless, we will review all the presented concepts in a unified and formal manner before finalizing this Chapter.

2.2 - Sharing the channel: the big problem in WiFi

WiFi networks may be defined as limited by interference. This means that the overall network performance highly depends on the interference among all the WiFi devices. Indeed, interference disturbs in the sense that if two radio signals reach a receiver at similar times, there is a collision and the frame is lost. A **frame** consists of a series of bits that carry information. Every frame has a **header** (with metadata about how to deliver the information) and a **payload** or **body** (with the information itself from higher layers). A frame is the basic unit of the *MAC (Medium Access Control)* layer, an entity present in every communication system that ensures the proper reception of transmitted data, as well as device identification and capabilities discovery.

2.2.1 – Acknowledge the information

Most of the frames transmitted in WiFi are <u>Data Frames</u>, long series of bits with user information encapsulated in the payload. However, they will not be successfully delivered without the help of other useful frames. Figure 4 shows the structure of a Data Frame in the 802.11 standards.

Bytes	2	2	6	6	6	2	6	0-2312	4
Data Frame	Frame Control	Duration	Address 1	Address 2	Address 3	Sequence Control	Address 4	**Frame Body**	CRC

Figure 4. Data Frame structure in 802.11.

The payload (data from higher layers in the protocol stack) is carried within the Frame Body (with variable length). The Control Frame field contains a series of bits that specify the subtype of the frame and additional information. The Duration specifies the length of the frame. The *CRC (Cyclic Redundancy Check)* provides redundancy to protect the data from noise and interferences. The Sequence Control field helps the link to keep the whole fragmented data stream in the intended order. Some of these fields will appear in other frame types.

The reader may have noted that there are four MAC addresses in a Data Frame. **A MAC address consists of 6 Bytes univocally identifying a physical interface** to transmit or receive bits. The general rule for the normal operation of a WiFi network states that for the structure of a Data Frame:

- Address 1 indicates the **receiver** of a frame. Known as *DA (Destination Address)*. The AP in the uplink, and the STA in the downlink.

- Address 2 indicates the **transmitter**. Known as *SA (Source Address)*. The AP in the downlink, and the STA in the uplink.

- Address 3 indicates the *BSSID (Basic Service Set Identifier)* of the network. That is, the MAC address of the AP that identifies that network. It may be the DA (in the uplink) or the SA (in the downlink). **This address helps the receiver to filter only the frames that belong to its network** (as we will see below, the radio channel is shared among all the near devices).

- Address 4 is present in more complex forms of communications. Rarely used.

As stated before, interference implies a high probability of frame collision. Fortunately, after the collision, **the transmitter is aware that it needs to retransmit the frame due to the absence** of an expected *acknowledgment (ACK)*, a short frame that is sent back to the transmitter after every success. It is interesting to note that the retransmission does not happen immediately: there is a back-off mechanism that minimizes the odds of a new collision.

But, what is exactly a collision? Why do frames collide? Below the MAC layer, the *PHY layer* of the transmitter builds a waveform by modulating the bits to be transmitted and sends it through one or more antennas. On the receiver, the reverse process is done: the PHY demodulates the bits and, if there was no error, the MAC layer receives the intended information. Figure 5 shows the transmission of a Data Frame in WiFi.

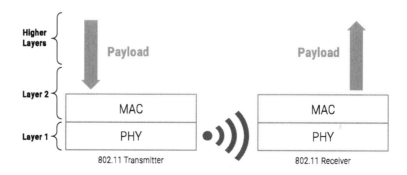

Figure 5. Basic 802.11 data transmission.

A collision occurs when the PHY is not able to recover the original bits, as multiple signals from uncoordinated transmitters have been added in the air (due to the linearity of Maxwell's equations). In this case, the received waveform does not correspond to any valid series of bits and thus the frame is discarded.

2.2.2 – The collision domain

The standard establishes two frequency bands, one around 2.4 GHz and another one around 5 GHz. This latter one has smaller coverage areas but achieves higher data rates within the room where the AP is located. These well-known bands are separated into various channels, and each network *BSS (Basic Service Set)* is deployed in one of those channels. A BSS is given by the coverage area of the *AP (Access Point)* and the *STAs* (user stations, like smartphones, tablets, and laptops) which are connected to it. **The collision domain is the combination of the channel and the BSS** because every device within it may implement a shared medium access mechanism. It is very surprising that the 802.11 standard only plans a few non-overlapping channels for all neighboring transmissions.

Traditionally, the assignment of the channel is done by the network administrator. It is selected manually for the deployment of the APs, and it remains static. However, modern wireless environments are subjected to changing conditions, specifically with interferences, due to novel network installations, mobile devices, and the proliferation of new electromagnetic sources. Obviously, this leads to inefficient use of the spectrum and an unacceptable rate of collision within the multiple access framework. Indeed, **several isolate networks are operating on the same collision domain**, and this is only due to a poor network management and channel assignment. Figure 6 shows a real-world spectrum of the 2.4 GHz band, with dozens of uncoordinated networks fighting for the control of the collision domain.

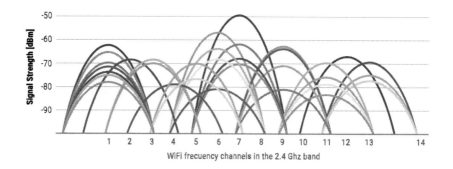

Figure 6. A typical WiFi residential crowded spectrum.

2.2.3 – Ask permission before speaking

In order to minimize the number of collisions, some rules to transmit were established. The shared media access method[6] in 802.11 is known as *CSMA-CA (Carrier Sense Multiple Access – Collision Avoidance)*. It is built upon the ideas of Ethernet access method, known as *CSMA-CD (Carrier Sense Multiple Access – Collision Detection)* that listens to the channel before transmitting. The problem with CSMA-CD is the following: in wired channels, **the transmitter knows whether the receiver is busy or not, because they share a common wired bus.** However, in wireless channels, one cannot listen to the radio availability at the receiver unless it probes the receiver itself with an *RTS (Request To Send)* frame. The receiver responses with a broadcast *CTS (Clear To Send)* and the rest of the users that are listening must stay quiet.

This "ask permission before speaking" process leads to the famous CSMA-CA method[7] which is implemented in each 802.11 channel. In Figure 7, the basics of CSMA-CA operation are shown. Usually, user stations communicate exclusively with the AP, **a fixed WiFi device which connects wireless devices**

---

[6] *Verdu, Sergio. Multiuser detection. Cambridge university press, 1998.*

[7] *Tanenbaum, Andrew S. "Computer networks, 4-th edition." ed: Prentice Hall (2003).*

to the wired part of the network. Some APs may have routing capabilities (e.g. those that most of us have at home), but others may not (they just act as a wireless hub). APs are connected to the Internet by Ethernet cables, hubs, switches, and routers.

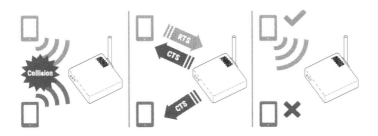

Figure 7. The CSMA-CA solution to the problem of shared media access.

Fortunately, as the WiFi technology only cares about what happens on the RAN (and only at low level), in this book we will not address wired infrastructures with these devices. In the following, will analyze only communications between APs and STAs. **As they are critical, these frames are sent at maximum transmitted power, minimum data rate, and maximum redundancy to maximize the odds of a successful reception.** This conservative mode of transmission is known as *legacy*. The original purpose of RTS/CTS was basically a way for 802.11g stations to **tell 802.11b stations to be quiet**. Since 802.11b stations can't understand the OFDM transmissions, the new stations have to say to 802.11b stations "please don't interrupt while I talk in a foreign language."

2.2.4 – Structure of Control Frames

The presented ACK, RTS, and CTS frames are the three more important frames of the <u>Control Frames</u> type. There are other types of Control Frames:

- *Block ACK*. Used to **acknowledge a block of Data** Frames. It saves time by acknowledging a set of Data Frames, without sending individual ACKs. This feature was first introduced in 802.11n.

- *BAR (Block ACK Request).* Sent by an STA that wants to check whether the receiver is capable of sending Block ACKs.

- *PS-Poll (Power Save-Poll).* Sent by an STA that is waiting for more data.

- *CF-End (Contention Free-End).* A contention-free period ends. Rarely used.

Note that this collision avoidance mechanism may be deactivated, removing the need for RTS/CTS frames but increasing the number of collisions. ACK frames are mandatory as the transmission stream does not continue without the proper acknowledgment from the receiver. **Control Frames are simple and effective**: unlike Data Frames, they do not have Frame Body nor Sequence Control. The main Control Frames are shown in Figure 8:

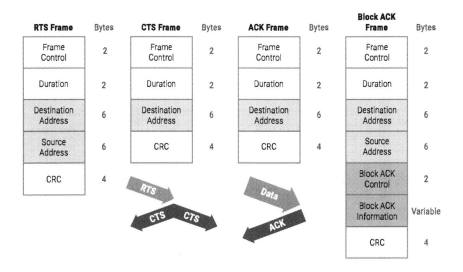

Figure 8. The most used Control Frames in 802.11. CTS appear as responses to RTS, and ACKs appear acknowledging success data reception.

The Frame Control, Duration, and CRC fields should be familiar from the Data Frame Structure of Figure 4. In addition, there may be **one or two addresses**:

- Frames **where the transmitter address is obvious or irrelevant** (CTS and ACK), only includes the DA, so the receiver pays attention to that frame. The transmitter address is obvious at the receiver because these Control Frames are response to other (RTS and Data, respectively) frames. The receiver remains waiting for those response frames from a specific transmitter.

- Frames **where the transmitter is not obvious and relevant** (RTS and Block ACK), includes the DA and the SA. This helps the receiver to identify from which device the frame is coming. Note that Block ACK frames have additional fields to set the type of blocks, the acknowledgment policy, etc.

2.3 - Aware of the environment: is anyone out there?

The previous Section has introduced Data and Control Frames, in the context of a network with only one AP and several STAs. However, in network deployments with multiple Access Points, there may be more than one AP candidate to serve an STA. In this Section, we introduce the last type of 802.11 frames: _Management Frames_. **These frames assist the discovery and association of devices,** so there is a mandatory interchange of Management Frames[8] before APs and STAs are ready to send and receive Data Frames. They have the structure shown in Figure 9.

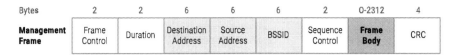

Bytes	2	2	6	6	6	2	0-2312	4
Management Frame	Frame Control	Duration	Destination Address	Source Address	BSSID	Sequence Control	**Frame Body**	CRC

Figure 9. The 802.11 Management Frame structure.

---

[8] D. Akin, and J. Geier, "CWAP - certified wireless analysis professional official study guide", Mc.Graw-Hill, 2004.

They are a bit more complicated than "Control Frames" presented in the previous Section. Note that **most of the Management Frames are acknowledged by the receiver**.

2.3.1 – Scanning your surroundings

According to the 802.11 protocol, the STA decides to which AP it authenticates and associates with. But before that, **it has to be aware of the available APs in its environment**. When an STA wants to join a new WiFi network, or when it wants to roam to another zone, it scans to see which ones are available in its surroundings. At the time of the scan, there are 2 possibilities:

- **Passive scan:** The APs periodically send (typically 10 times per second) "*Beacon*" frames. These beacons are used to announce their presence and to facilitate its detection by potential STAs in the area. It includes an *SSID (Service Set Identification)*, which is the typical name of the WiFi network that every user sees. The simile **would be a lighthouse that emits pulses so that the ships can notice its presence**. In passive scanning, the client devices listen to fresh beacon frames available in their surroundings. Then, STAs evaluate these frames and decide which network to connect. This process is slow since the STA has to tune to each frequency channel, and wait to receive a Beacon.

- **Active scan – Probing:** This process aims to accelerate the previously described process of passive scanning. To achieve that, the standard proposes the "*Probe Request*" frame. Indeed, when an STA wants to connect to a new network, instead of waiting to receive the beacon, it sends a Probe Request frame on each frequency channel. **A simile would be to enter a dark cave asking if anyone is there**. When an AP receives a "Probe Request" it answers with a "*Probe Response*" frame indicating its presence. The STA then decides to which one wants to connect. Indeed, STAs may request to connect to a specific WiFi network (already known) by setting the SSID within the Probe Request Frame info.

2.3.2 – Binding the devices

Whatever it uses active or passive probing, the STA decides to which one to send its "*Authentication Request*" frame. This frame triggers the binding process between AP and STA. If AP determines that the STA has the credentials, it sends back an "*Authentication Response*" frame. Otherwise, it will send a "*Deauthentication*" frame, returning to the initial state. Note that **an STA can be authenticated to multiple APs however it can only be actively associated with a single AP at a time**. Figure 10 summarizes this process with both passive and active scan (ACKs are omitted).

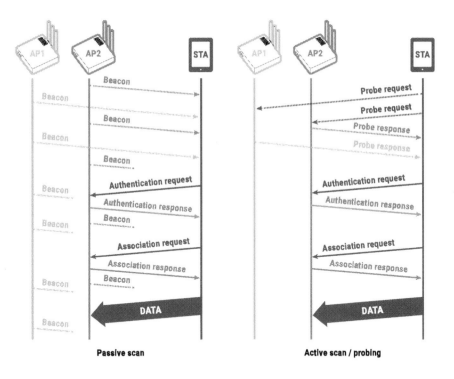

Figure 10. Passive (left) versus active (right) scanning paradigms when discovering and connecting to a wireless network.

At this moment, the STA has been authenticated but not yet associated. Once a mobile STA determines which AP it would like to associate to, it will send an "*Association Request*" frame to that AP. This frame contains chosen encryption types (if required) and other compatible 802.11 capabilities. If it matches the capabilities of the AP, it will answer with an "*Association Response*" frame, with a success message granting network access to the STA. **Now the link has been successfully established and data transfer can begin.** It the capabilities does not match, the AP sends a "*Disassociation*" frame, going back to the previous state.

2.3.3 – Specific management parameters

Each of these Management Frames has its own Frame Body parameters. In the following, we give an overview of some of the most important ones:

- **Beacon and Probe Response**: timestamp (to synchronize clocks), beacon interval (time between beacons), capability information, **SSID**, supported rates, physical parameter sets, and traffic indication map (to indicate to any sleeping STA that the AP has buffered frames for it).

- **Probe Request**: SSID and supported rates.

- **Authentication Request and Response**: authentication algorithm, transaction sequence number, status code, and challenge text (for some types).

- **Association Request**: capability information, listen interval (how often an STA in power save mode wakes to listen), SSID, and supported rates.

- **Association Response**: capability information, status code, association ID, and supported rates.

- **Deauthentication and Disassociation**: Status code.

In addition, there are other rarely used Management Frames like *"Reassociation Request"* and *"Reassociation Response"* frames (to transition to another AP of the same network), or *"Action"* frames (to trigger specific actions in the cell, like radio measurements or fast transitions). Action frames are gaining more attention in the last years as they provide new features regarding channel probing (See Subsection 5.3.2).

2.3.4 – Backward compatibility

It is important to note that this has been a very basic overview of the most important 802.11 MAC frame structures. As the standard evolves and becomes more complex, new capabilities must be introduced into the old frame formats. The question is: **How to establish new frame fields and parameters in existing 802.11, so that legacy devices keep understanding the classic frames, and new devices are aware of the new features of advanced standards?**. For both MAC and PHY layers, the answer is in the Frame Body and Data fields (respectively). These fields usually encapsulate the information from higher layers.

As we will see later, WiFi standards like 802.11ac and 802.11ax introduce several new features that only modern devices may understand and exploit. The Frame Body field is responsible for containing the required new parameters for an accurate operation.

- A **legacy** device that receives a modern frame will focus on only the part of the frame that it understands (the frame header). The part that it does not understand is within the Frame Body and treated as higher-layer information not intended for it.

- A **modern** device that receives a modern frame, is able to interpret and use these additional fields coded within the Frame Body.

In Figure 11, the structure for 802.11a/g (legacy) and 802.11ac (new) MAC Data Frames is compared.

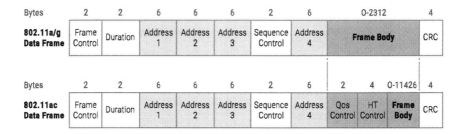

Figure 11. Legacy vs. new Data Frames.

HT stands for High Throughput (usually related to 802.11n), while VHT stands for Very High Throughput (usually related to 802.11ac). There are two new fields:

- The **QoS control field** comprises traffic identifier (voice, video, best effort, or background), EoSP (End-of-Service Period, so the client can go back to sleep), and ACK policy, among other parameters.

- The **HT Control field** contains information about link adaptation and calibration, channel probing, Null Data Packets announcement, and other 802.11ac features.

These elements will appear in advanced Chapters of this book. In addition, Management Frames in 802.11n and 802.11ac has additional fields to describe its advanced features. However, we will not analyze most of them. As we will see in the next Section, the same philosophy applies to backward compatibility in PHY frames (PPDU).

2.4 – Down into the physical layer

2.4.1 – Physical frames: Physical Protocol Data Units

The PHY frame is also known as *PPDU (Physical Protocol Data Unit)*. It can be seen as a **series of bits preparing and protecting the MAC frames for a**

successful transmission and reception on the radio channel. Figure 12 shows the structure of a PPDU. These bits will be modulated into symbols and transmitted by the radio modules through the antennas. In this book, **we will not deeply analyze non-OFDM frames and modulations**, as they are in disuse and not related to 802.11ax. Non-OFDM frames use the old DSSS modulation and are supported only by 802.11b/g compatible devices.

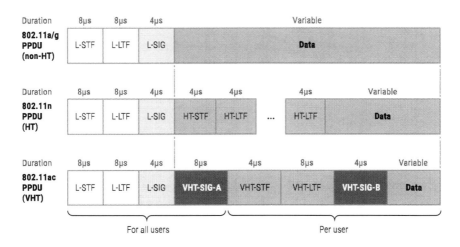

Figure 12. PPDU structure.

The **Data field** bears the frame from the MAC layer or even an aggregate frame with multiple MAC frames. It is important to remark that **if no Data field is present in the PHY payload, an *NDP (Null Data Packet)* is sent**. When a device's PHY layer receives an NDP, there is nothing to report to the MAC layer. As we will see later in advanced Chapters, this type of transmission (with only header information) is used in 802.11ac to perform advanced features like beamforming.

Regarding the 802.11ac PPDU header, the most important legacy fields are:

- *L-STF (Legacy Short Training Field)* and *L-LTF (Legacy Long Training Field)*: they are understood by legacy 802.11a/g devices (as well as modern devices), and consists on a series of 12 OFDM symbols that helps the receiver to align and synchronize with the incoming frame, as well as selecting the antenna. They are used for the start of packet detection and automatic gain control setting, as well as frequency offset estimation, channel estimation, and time synchronization.

- **L-SIG (Legacy Signal)** field: describes the data rate and length (and thus the duration) of the frame. Modern devices send this field in slow legacy rates (6 Mbps) to help the synchronization. It is used to set the data rate and MCS.

Figure 13 shows the amplitude and phase versus time of the symbols corresponding to the legacy preambles.

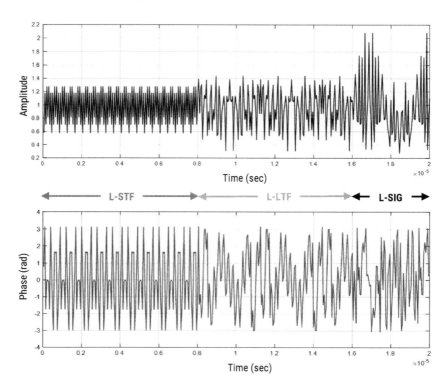

Figure 13. Amplitude (up) and phase (down) of the legacy preambles.

The next preambles are non-legacy.

- **VHT-STF (Very High Throughput Short Training Field)** and **VHT-LTF (Very High Throughput Short Training Field)**: As its legacy equivalent fields, they help the receiver to tune-in the signal, setting the gain. In addition, they help to demodulate the rest of the frame depending on the number of streams, to estimate the channel for beamforming, and other advanced features. They have a repeating pattern similar to the legacy ones.

- **VHT-SIG-A (Very High Throughput Signal A)** and **VHT-SIG-B (Very High Throughput Signal B)**: These signals are only understood by 802.11ac devices. This two fields together describe the frame physical attributes (channel bandwidth and MCS, among others). They also mark whether the frame corresponds to a multi-user transmission.

Figure 14 shows the non-legacy (VHT) preambles:

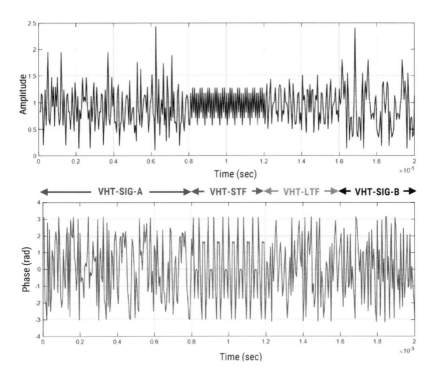

Figure 14. Amplitude (up) and phase (down) of the non-legacy (VHT) preambles.

Finally, the **Data field** is appended after the preambles. Figure 15 shows the complete waveform of an 802.11ac frame (Data field is much longer than shown).

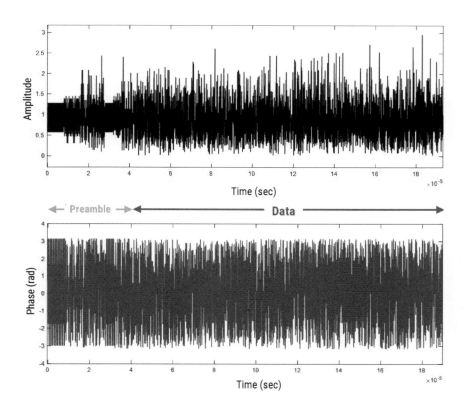

Figure 15. Amplitude (up) and phase (down) of complete frame waveform. Note that the Data field is much longer than shown in this figure.

2.4.2 – Transmission and reception

Before analyzing the main features of the PHY layer, this Subsection gives an overview of the main blocks in a transmitter-receiver chain for WiFi devices. In Figure 16, we will only analyze the most important subsystems that will be further addressed in 802.11ax.

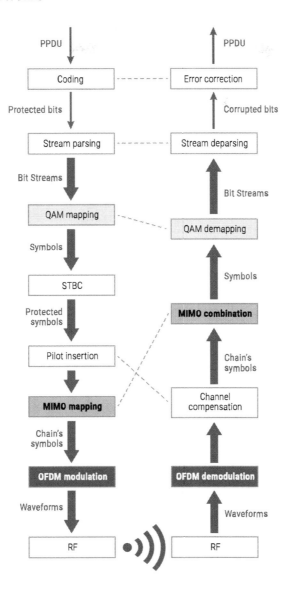

Figure 16. The main structure of WiFi transmitters (left) and receivers (right). The colored subsystems will be further analyzed in this Section. Corresponding transmitter-receiver subsystems are related by a dashed line (although they are not physically connected). Finally, the transformed information between subsystems is written above most of the links.

The most important processes in a modern WiFi <u>transmitter</u> chain are:

- **Coding:** after scrambling the bits, they are passed through a convolutional or *LDPC (Low-Density Parity Check)* coder[9], in order to add redundancy and protect the data from the channel noise and interferences.

- **Stream parsing:** divides the bits into spatial streams if the devices have more than one antenna. These streams may, or may not, correspond to physical antennas.

- **QAM mapping:** bits are mapped onto QAM constellation[10] points using the desired modulation (4, 16, 64, or 256-QAM). The new 1024-QAM will be introduced with 802.11ax. The QAM modulation will be explained in the next Subsection.

- **STBC[11] (Space–Time Block Coding):** an optional step to add redundancy. It consists of sending one spatial stream across multiple antennas.

- **Pilot insertion:** pilot symbols are known symbols at the receiver, used to estimate the channel and to align parameters between both sides of the communication link. Transmission data and pilots are combined to create the complete data set.

- **MIMO mapping:** Spatial streams (maybe after STBC) are mapped into the different transmit chains (one per antenna). It may be a *direct mapping*, or a *spatial expansion* (one for all the antennas), like in beamforming.

---

[9] *Gallager, Robert. "Low-density parity-check codes." IRE Transactions on information theory 8.1 (1962): 21-28.*

[10] *Webb, William T., and Lajos Hanzo. Modern Quadrature Amplitude Modulation: Principles and applications for fixed and wireless channels. John Wiley, 1994.*

[11] *Tarokh, Vahid, Hamid Jafarkhani, and A. Robert Calderbank. "Space-time block codes from orthogonal designs." IEEE Transactions on information theory 45.5 (1999): 1456-1467.*

- **OFDM modulation:** an *IFFT (Inverse Fast Fourier Transform)* over the symbols builds an OFDM waveform which is ready to be transmitted.

- *RF (Radiofrequency):* the baseband signal is modulated to the selected channel. The waveform is then amplified to the set transmitted power.

On the <u>receiver</u> side, the chain follows up the dual structure.

- **RF:** a low noise amplifier raises the signal level on each receiver chain, and the mixers demodulate the waveforms to baseband.

- **OFDM demodulation:** an *FFT (Fast Fourier Transform)* recovers the symbols on each OFDM sub-channel[12]. OFDM will be explained in Subsection 2.4.4.

- **Channel compensation:** the preamble helps the receiver to compensate for frequency fading (sub-channels with different amplitudes due to propagation effects) in the OFDM spectrum. This is key to recover the original symbols without distortion.

- **MIMO combination:** once the symbols have been recovered, all the symbols are combined. If STBC was used, the complementary step is done.

- **QAM de-mapping:** The estimated bits are recovered by projecting the QAM received symbols into a decision alphabet. More on this on Subsection 2.4.3.

- **Stream de-parsing:** when MIMO is used only to add redundancy or beamforming, the original spatial streams must be recovered.

- **Error correction:** The last step is to recover/detect corrupted bits by using the redundancy introduced by the channel codes in the transmission chain. Longer codes are more powerful, but they result in less spectral efficiency.

---

[12] Brigham, E. Oran, and E. Oran Brigham. *The fast Fourier transform and its applications. Vol. 1. Englewood Cliffs, NJ: prentice Hall, 1988.*

The received PPDU will be (hopefully) identical to the transmitted one. If the error correction block detects corrupted bits (but it is not able to recover the original data), the **PPDU is discarded and the MAC layer is aware that it must not answer with an ACK to the transmitter.**

The next Subsection explains the main features of the PHY layer from a signal processing[13] point of view. These features will be expanded in the next Chapters, as they play a key role in 802.11ax.

2.4.3 – Quadrature-Amplitude Modulation

Depending on the modulation used, the transmitted signal bears the desired information in one way or another. In WiFi, QAM signal uses two orthogonal *basic transmission pulses* **to transport the bit stream in the shape of symbols.** A *constellation* is the representation (See Figure 17) of a waveform in the signal space[14], which is given by a basis (usually orthogonal).

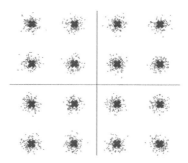

Figure 17. Received constellation (grey small dots) and constellation alphabet (dark × symbols) of a 16-QAM signal, used in the 802.11n/ac/ax standard with good conditions.

---

[13] Perahia, Eldad, and Robert Stacey. "*Next generation wireless LANs: 802.11 n and 802.11 ac*". Cambridge university press, 2013.

[14] B. Sklar, *Digital Communications: Fundamentals and Applications (2nd ed.), Prentice Hall (2001).*

Orthogonality helps to recover the desired symbols at the receiver when they are contaminated with noise. Noisy and contaminated signals are characterized by spreading dots in the signal space, and thus the received constellation shows clouds of symbols. The relation between signal and noise + interference powers is the *Signal to Noise and Interference Ratio (SNIR)*, and it is a key metric to measure the quality of the wireless link.

The set of available symbols to transmit is known as constellation *alphabet*, and **it may be seen as a template, which may help the receiver to successfully recover the information sent.** Figure 18 shows some basic QAM constellations in the signal space, for different modulation orders (the size of the alphabet): 4, 16, 64, and 256. In WiFi, the order of the modulation defines the MCS value, together with other PHY parameters (specifically, the coding rate and the number of streams).

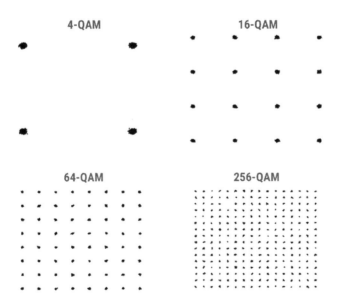

Figure 18. QAM modulations are widely used in the 802.11 standards.

For a fixed average (or maximum) transmit power, the bigger the modulation order is, the closer the symbols in the constellation are. This is the reason why in WiFi, high-order modulations (and thus MCS) are only allowed with good signal conditions (usually for STAs near the AP). At the receiver, the noisy signals are passed through a *detector* which decides which symbols were transmitted originally, based on proximity criteria.

As the reader may note, the closer the symbols are, the harder is to decide which symbol of the alphabet corresponds to the received noisy symbol. This implies that the *Bit Error Rate (BER)* grows with low SNIR. Figure 19 shows the BER vs SNIR (in dB) curves for the previous constellations. The best points of operation, in terms of efficiency, are the closest to the lower left corner of the figure.

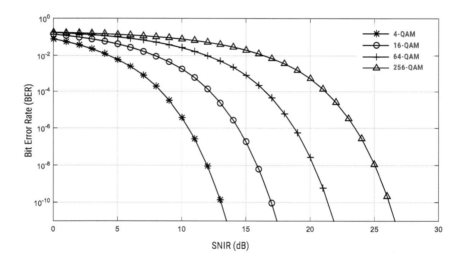

Figure 19. BER vs SNIR (in dB) for fixed transmit power and different constellations. The bigger the modulation order is, the closer the symbols in the constellation are.

2.4.4 – Orthogonal Frequency Division Multiplexing

Frequency Division Multiplexing (FDM) has been widely used in wireless and wired communications during the last century. It consists of modulating the frequency so the information is transmitted **at the same time over several frequency** channels.

Orthogonal Frequency Division Multiplexing (OFDM) consists of several narrowband subcarriers that transmit parallel (in frequency) streams of information. Each subcarrier is orthogonal to each other so the separation may be done at the receiver.

OFDM has been **the most successful modulation in the last decade** due to the increase in hardware capacity of signal processors. Its digital implementation is done by using the FFT (Fast Fourier Transform) on the receiver, and the IFFT (Inverse Fast Fourier Transform) on the transmitter. A *Cyclic Prefix (CP)* is added to accelerate the modulation and to avoid spurious spectral tails. These algebraic operations modulate the QAM symbols into subcarriers, generating a waveform that is ready to be transmitted by the radio chains of each antenna.

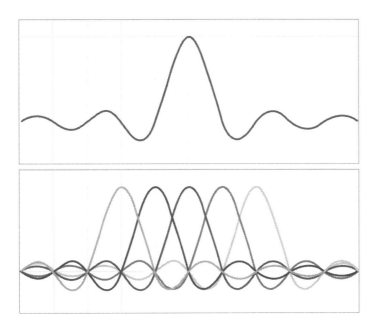

Figure 20. (Upper) One subcarrier. (Lower) 5 orthogonal subcarriers builds and OFDM.

Figure 20 shows the spectral shape of one subcarrier and five subcarriers, forming an OFDM. We will not delve into the mathematical details of OFDM, but it is important to highlight that it has some benefits especially for indoor wireless communications:

- As a different data stream is modulated into each narrow subcarrier, frequency-selective channels are compensated by just adjusting the original amplitude of each subcarrier.

- As the subcarriers are orthogonal, the channels are narrower than usual and thus the spectral efficiency is improved (more channels in the same bandwidth).

- FFT and IFFT operations are easily done nowadays in commercial WiFi chips.

- Some subcarriers bear specific data known as *pilots*, which are known at both transmitter and receiver sides. **Pilots are used to probe the radio channel characteristics, in order to adapt the receiver chain** to dynamic channel events (moving devices, unpredictable propagation phenomena, etc.).

- A typical 802.11a/g channel has 20 MHz bandwidth, although in 802.11n/ac amendments it may be enlarged to 40, 80, and even 160 MHz. Other rare amendments allow 10 and even 5 MHz channels, but they are very rare.

- In 802.11a/g channels there are 52 subcarriers, numbered from -26 to +26. 48 of them bears data with a QAM modulation, and 4 of them (numbers -21, -7, 7, and 21) bears pilot symbols with a BPSK modulation. The central subcarrier (number 0) is null. Subcarriers are spaced 312.5 kHz each.

- 802.11n/ac channels have more subcarriers, depending on the bandwidth:

    o   20 MHz channels have 56 subcarriers (48 with data and 4 with pilots).

    o   40 MHz channels have 114 subcarriers (108 with data and 6 with pilots).

o   80 MHz channels have 242 subcarriers (234 with data and 8 with pilots).

o   160 MHz channels have 484 subcarriers (468 with data and 16 with pilots).

The modern WiFi OFDM structure is shown in Figure 21.

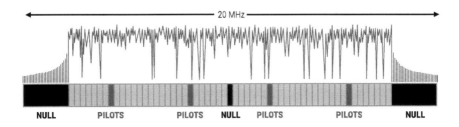

Figure 21. 802.11n/ac OFDM structure and power spectral density.

It is interesting to note that, when a device transmits, it occupies all the channel for which the BSS has been configured. **Only one device can transmit at a time**. As we will see later, this is the main change in 802.11ax.

2.4.5 – Multi-antenna communications

In signal processing, a communication system with multiple antennas is known as *MIMO (Multiple-Input, Multiple-Output)*. Since the 802.11n standards, multiple antennas have been incorporated into WiFi devices for various purposes. In the following, we will give a brief overview of the use of MIMO in modern wireless systems:

• To enhance **reliability**, by transmitting the same or redundant information from different antennas, and combining them at the receiver. This is known as *STBC (Space-Time Block Coding)* or *MRRC (Maximal Ratio Receive Combining)*. It does not need more than one antenna at the receiver, although it can exploit them.

- To increase **coverage area**, by steering the radiation beam of an array of antennas in a specific direction. This is known as *BF (beamforming)* and needs the feedback of the channel coefficients from the receiver. It does not need more than one antenna at the receiver (typical in WiFi downlink).

- To maximize **throughput**, by creating virtual streams between pairs of transmitting and receiving antennas. When there are only one transmitter and one receiver, both with multiple antennas, this is known as *Spatial Division Multiplexing (SDM)*. When there are several devices using this method to access the shared channel, this is known as *Spatial Division Multiple Access (SDMA)*. It needs a receiver with more than one antennas (only high-end WiFi STAs have more than one antenna at the receiver).

- To **divide the collision domain** by simultaneously steering different beams to different STAs. This is known as *Multi-User MIMO (MU-MIMO)*, which has been successfully implemented in 802.11ac-Wave2 (downlink) and 802.11ax (uplink). It does not need more than one receiver antenna.

Figure 22 shows the difference between beamforming and SDM in terms of quality of signal reception. A 2x2 MIMO system with beamforming improves the SNIR, so the received symbols may be demodulated with less error.

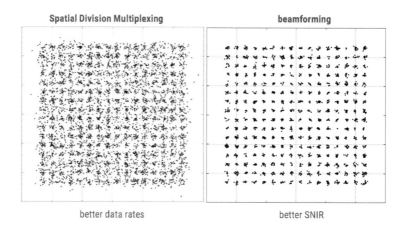

Figure 22. Comparison between SDM (left) and beamforming (right) symbols.

Table 3 summarizes the aforementioned use of MIMO for WiFi communications. Note that there may be special cases or interpretations where the table is not exact, but it aims to give a general overview.

	Introduced in standard	Aim	Error rates with interferences	Single STA data rates	Aggregated STAs data rates	Signal strength	Needs more than 1 antenna in receiver?
STBC	802.11n	Adds redundancy	↓	=	=	=	NO, but helps
SDM	802.11n	More spatial streams	=	↑	=	=	YES
Beamforming	802.11n	Increase range	↓ due to higher signal strength	=	=	↑ for a fixed position	NO
MU-MIMO	802.11ac Wave 2	Divide collision domain	=	=	↑ cach STA gets the maximum	↑ for a fixed position	NO

Table 3. Overview of MIMO features.

802.11ax will introduce some of this uses in the uplink, for which it will describe a scheduling protocol for the first time in WiFi history.

2.5 – Conclusions

In this Chapter, we did an overview of the building blocks of the 802.11 standards, from the justification of **each management and control frame**, to the physical features behind MIMO and OFDM. As this is not an introductory WiFi book, we deliberately did not explain each frame structure and data field.

The reader may have now a complete understanding of the main features of the standard, up to 802.11ac. The next Chapter introduces 802.ax. By using this Chapter as a basis, the present and future challenges for WiFi, as well as the improvements proposed by 802.11ax, shall be clear.

3

An overview of 802.11ax

3.1 - Introduction

This Chapter introduces the main problems of modern WiFi networks, as the exponential growth of devices and transmitted data. We also review the main issues of electromagnetic wave **propagation in indoor/outdoor and mobile environments**. Finally, we overview the main features of 802.11ax with respect to the 802.11ac standard. Most of these features pursue a concurrent transmission and reception of data and a more efficient use of radio resources.

3.2 - Problems in modern WiFi networks

In this Section, we overview some of the serious problems that WiFi devices have been issuing in the last two decades. Some of them are inherent to the radio channel, while others were early adopted for the ease of the operation. As we will see later, the new amendment 802.11ax aims to solve most of them by introducing advanced features that have been possible mainly thanks to **hardware improvements in the last years**. Figure 23 shows some numbers of today's WiFi technology. These numbers will be addressed in the next Subsections as the main pillars of WiFi challenges for future of wireless communications.

3.2.1 – Shared access to the transmission media

The asynchronous usage of the electromagnetic spectrum was proposed for 802.11 in the late 90s due to its simplicity and ease of implementation. However, it belongs to another century, and nowadays it is **operating in a world for what it was not designed**. Indeed, there are only a few practical channels in the 2.4 and 5 GHz bands.

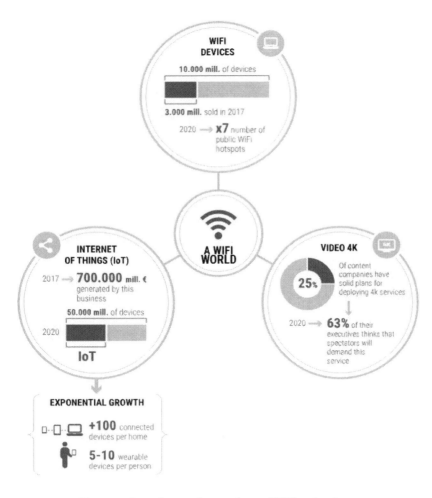

Figure 23. Some impressive numbers of WiFi technology.

In addition, WiFi shares the unlicensed spectrum with other communication and exogenous technologies like Bluetooth, Zigbee, cordless phones, microwave ovens, baby monitors, etc. The CSMA/CA method is prepared for high-interference bands like these ones, but it results in an excessive inefficiency in the use of radio resources.

Due to backward compatibility issues, 802.11 designers decided not to schedule (as in mobile communications, like 3G, 4G, or LTE) the access to the shared media ... until 802.11ax. In the next Chapters, we will show how the standard is prepared for the next step. **For the first time in WiFi, transmissions will be scheduled** and orchestrated by a coordination entity.

3.2.2 – The high density of users and things

In 2018, **there are more wireless devices in use than humans on the planet**[15]. The number is near 10,000 million and keeps growing at a high rate, with more than one-third of those devices[16] being *M2M (Machine to Machine)*, without user connection. More than 3,000 WiFi devices were sold in 2017, and the services built around wireless communications generated the 4.4% of the global *GDP (Gross Domestic Product)*. From the infrastructure point of view, in 2020, the number of public WiFi hotspots will fold 7x, reaching more than 430 million.

In addition, this trend is going to grow exponentially in the next years, thanks to the irruption of *IoT (Internet of Things)*. The business around this new paradigm has reached 700,000 million € at the end of 2017, and it is estimated[17] that **there will be more than 50,000 million devices connected to the internet in 2020.** Half of them will correspond to IoT. A great portion of them

---

[15] *https://www.gartner.com/newsroom/id/3598917*

[16] *https://emear.thecisconetwork.com/site/content/lang/en/id/3171*

[17] *https://iot.telefonica.com/multimedia-resources/infographic-iot-trends-for-2017*

will access through WiFi, thanks to the advantages of the 802.11 standards against Bluetooth, Zigbee, LTE (Long Term Evolution), and similar wireless technologies. The most conservative forecasts consider hundreds of connected devices for the average home, and each one of us shall wear 5-10 devices with a wireless chip integrated.

3.2.3 — High demand for multimedia traffic

Furthermore, the rise of requested data rates (throughput) supposes a bottleneck for current RANs (Radio Access Networks), demanding the use of new technologies like MU-MIMO and OFDMA. Those enabling technologies will be explained in depth later in this book.

Cisco confirmed that the overall WiFi traffic from both mobile devices and WiFi-only devices together will account for **almost half (49%) of total IP traffic by 2020,** up from 42% in 2015[18]. The rise of transmission rates is given by the new services demanded by users, mainly real-time video in downlink and uplink directions. Of course, resolutions above FullHD (2 Mpixels per frame) and 4K (8 Mpixels per frame) are expected. **This implies data rates of dozens of Mbps for each one of the users** connected to a WiFi infrastructure. Nowadays, a quarter of main content-creation companies[19] have solid plans to launch 4K services, specifically on-demand video for Smart TVs, tablets, and smartphones. For that reason, 63% of executives of those companies bet on that those resolutions will be mandatory for the mainstream before 2020. Furthermore, YouTube users upload 48 hours of new video every minute of the

---

[18]*https://www.cisco.com/c/en/us/solutions/collateral/service-provider/visual-networking-index-vni/mobile-white-paper-c11-520862.pdf*

[19]*http://www.intelsat.com/wp-content/uploads/2014/09/Fast-Forward-to-4K-future-infograhic.jpg*

day[20]. In addition, video-gaming and industrial sectors demand latencies below very severe limits.

In addition to high-resolution video streaming, there is an increasing demand for reliable and fast WiFi communications for the video gaming sector, where latencies and jitter must remain below very severe limits. Finally, the social networks have created the demand for fast uplink communications on WiFi users. They want to instantaneously share photos and videos with their contacts, so **robust STA-to-AP transmissions is one big issue that 802.11ax aims to solve.**

The era of information gives a step to the era of attention. Each year, humans and things are generating more digital data than ever before (90% world's historical data has been created in the last 2 years), and most of this information is not exploited by the main players. Between early 2018 and 2020, **it is expected that the digital data will double**[21]. Network owners are very interested in applying analysis to the gathered information, but they need to have access to fresh and secure data from the network.

3.2.4 – Complex propagation phenomena

The new standard 802.11ax is going to deal with propagation issues in modern wireless communications[22]. Nowadays, most of the wireless communications occur in cities, with mobile devices and strong electromagnetic interference. In addition, the geometry of buildings yields to multiple reflections of the transmitted wave.

---

[20] https://fortunelords.com/youtube-statistics/

[21] https://www.emc.com/leadership/digital-universe/2014iview/executive-summary.htm

[22] Goldsmith, Andrea. Wireless communications. Cambridge University Press, 2005.

Firstly, *Line-of-Sight (LoS)* communications exhibit a better performance than *NLoS (Non-Line-of-Sight)*. **In LoS, there is an obstacle-free path between transmitter and receiver.** In NLoS, the signal goes through obstacles, resulting in reflection, refraction, diffraction, and absorption phenomena. The typical result for a brick wall is a drop on the signal strength of about 5 dB, depending on the type of material. This is about dividing by four the received watts (a 3 dB drop halves the signal strength). For that reason, NLoS attenuation is one of the main problems in WiFi indoor deployments.

It is interesting to remark that attenuation affects more severely to 5 GHz channels. In general, **higher frequencies are worse for NLoS communications,** although they are likely to be clean of interference. Obstacle absorption coefficients are lower on the 2.4 GHz band, an even lower on sub-GHz frequencies used in other communication standards like 3G, 4G, or 802.11ah. The golden rule for wireless indoor deployments is: *"the higher the frequency, the lower the coverage"*.

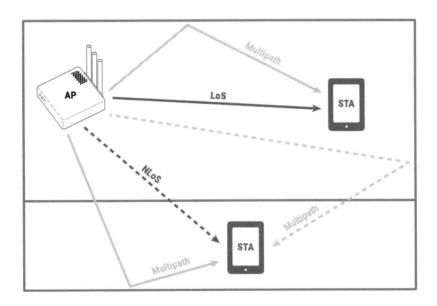

Figure 24. LoS, NLoS, and multipath propagation.

Secondly, the ray model for indoor communications establishes that **the received signal is a combination of the direct path (LoS or NLoS) and several indirect paths**.

This is the well-known *multipath propagation*, shown in Figure 24. The main implications of multipath propagation are that:

- On the time domain, the received waveform is a linear convolution of the transmitted waveform with the channel impulse response. **A signal mixed with its delayed and attenuated copies**. It the copies are delayed enough, one symbol may be confused with the next one. This is the dreaded *ISI (Inter-Symbol Interference)*, and 802.11 establishes a guard interval between OFDM symbols to deal with it.

- On the frequency domain, the effect is **a valley or a peak on the spectrum of the received signal**. Two or more sinusoidal signals may combine in an additive or subtractive manner, increasing or decreasing (and even nulling) the received waveform at specific frequencies.

Related to multipath, *fading* is one of the most destructive effects that an environment can do on your electromagnetic signal. It supposes **a considerable drop in the signal strength,** depending on some variables like time, frequency, or space. There are several types of fading (slow and fast, with different statistical models, etc.), related to different phenomena. The two most important for WiFi are:

- **Selective fading**: it is caused by multipath propagation, canceling one specific frequency on the signal spectrum. It is like a frequency filter. This may result in several bits lost (the ones carried on the canceled subcarrier). Fortunately, the OFDM modulation is designed to overcome this spectrum distortion by splitting the spectrum into several, narrow channels of 312.5 kHz. The signal may be equalized at reception by using the knowledge extracted from pilots. Figure 25 shows an example of frequency-selective fading.

- **Flat fading**: most of the signal spectrum is affected by the fading (in the same magnitude). It is related to moving objects between transmitter and receiver. It can block the communication link during several seconds.

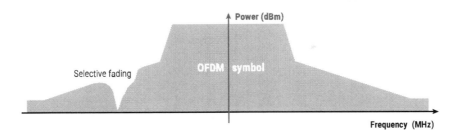

Figure 25. Frequency-selective fading

Finally, outdoor users are starting to request reliable and fast WiFi networks. *Outdoor communications* have been the exclusive territory of mobile networks like 3G, 4G, and LTE. However, the throughput and high-density adaptation of modern 802.11 standards shall not be reached by nowadays mobile standards (even 5G). In addition, **a WiFi network is cheaper to deploy, and the spectrum is unlicensed**. Nevertheless, outdoor environments present specific challenges for WiFi: larger and multiple propagation paths, deeper fading, strong attenuation, and high-density of users requesting high-bandwidth data (like stadiums). Figure 26 shows the effect of fading in the received spectrum of a WiFi signal.

It is important to note that, until now, WiFi has survived in these challenging scenarios. However, the performance and throughput of 802.11 devices were severely degraded due to the absence of standardized tools to deal with these issues. In general, the responsibility of providing a good connection is an issue for the network administrators and deployment team.

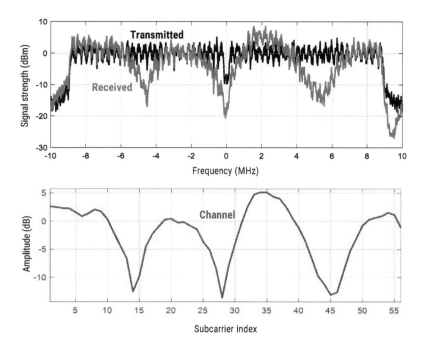

Figure 26. The spectrum of transmitted and received OFDM signals (up). Channel response (down) with frequency-selective fading.

3.2.5 – Energy consumption and green communications

Energy consumption of wireless devices is considered a major issue in today's communication systems. The increase in energy density of current state of the art (Lithium-Ion) batteries is far from following Moore's Law. **The current trend forecasts "just" a twofold increase in the next 10 years**[23]. Furthermore, battery-powered devices are becoming frequently part of the WiFi infrastructure itself. Not only handheld devices. This has a profound

---

[23] *Garcia-Saavedra, A., Serrano, P., Banchs, A., & Bianchi, G. (2012, December). Energy consumption anatomy of 802.11 devices and its implication on modeling and design. In Proceedings of the 8th international conference on Emerging networking experiments and technologies (pp. 169-180).*

impact on the design of IoT[24] and sensor networks. However, in the last years, it has been a huge research effort with the aim of reducing energy consumption. A wireless device has several sensors (IMU – Inertial Measurement Unit, light sensors, touch screen, cameras) and actuators (photo flash, 4K screens) that consume a lot of energy. Those related with wireless communications (Bluetooth, GPS – Global Positioning System, WiFi, NFC – Near Field Communications, and more) can have a great impact on power consumption of a laptop, a smartphone, or a tablet. Efficient medium access protocols and novel physical layer features will help to reduce considerably this issue. We can distinguish three different paradigms for WiFi networks:

- Until 2005, it can be considered that wireless networks were **coverage-dominated**, where expensive deployments where done to expand the operational range. Power consumption was not a worry for the standard.

- From 2005 to 2015, wireless networks were **capacity-dominated**, where all standardization efforts were in the direction of maximizing throughput and data rates for all the users. Achieving high throughput was more important than reducing radiation and power consumption. MIMO and wider channels were introduced to reach this objective.

- Since 2015, standardization bodies have started to pay attention to *greener communications*. We have entered into the **efficiency-dominated** era. In this philosophy, reducing energy consumption and radiation is the main focus, as achievable data rates of WiFi networks satisfying the demands.

As we will see later in this Chapter, 802.11ax brings new features for reducing power consumption and electromagnetic radiation, increasing efficiency and autonomy of wireless devices. In addition, those efforts also mitigate interference between BSS, pursuing a cooperative behavior.

---

[24] *Cui, Xiaoyi. "The internet of things." In Ethical Ripples of Creativity and Innovation, pp. 61-68. Palgrave Macmillan, London, 2016. Harvard.*

3.3 – OFDMA and resource planning

OFDM is a multi-carrier modulation method, where bits are modulated into narrow sub-carriers. At each time, one user occupies the channel with its waveform, and the rest of the transmitters must wait before obtaining the control of the channel. Unlike OFDM access, OFDMA[25] (Orthogonal Frequency Division Multiple Access) **allocates different blocks of subcarriers for different STAs at different time slots.** Figure 27 shows the difference.

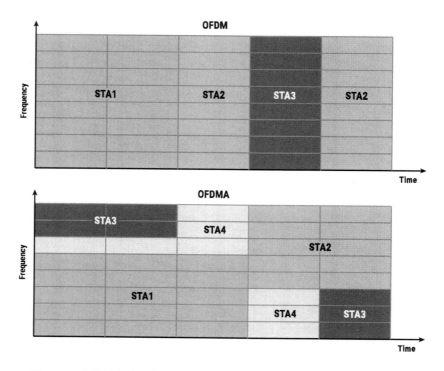

Figure 27. OFDM (up) and OFDMA (down) access in the time-frequency plane.

---

[25] *Seong, Kibeom, Mehdi Mohseni, and John M. Cioffi. "Optimal resource allocation for OFDMA downlink systems." Information Theory, 2006 IEEE International Symposium on. IEEE, 2006.*

OFDMA works in both uplink and downlink directions, allowing several devices to transmit or receive at the same time. This is a new paradigm for WiFi, where transmitted signals have been competing for the physical channel until now. However, the OFDMA access necessarily implies **a resource planning protocol**, to schedule all device transmissions. We will see later in this book the new frames that allow coordination of WiFi nodes for resource planning.

3.4 – Downlink and uplink MU-MIMO

Downlink MU-MIMO was introduced with 802.11ac (2013), but it has not been implemented until recently (2016) when Wave 2 devices came to the market. As stated in Section 1.4.4, MU-MIMO[26] allows an AP to steer different data streams to different STAs, by beamforming the energy of its antenna array. It implies a significant effort on channel estimation, feedback, and beamforming, so both transmitter and receiver must be compatible. Nowadays, there are only a few MU-MIMO smartphones, tablets, and laptops in the market, but they will become a commodity during 2018 and 2019.

However, **many STAs are single-antenna (smartphones)**. Furthermore, due to energy and battery constraints, even two antenna STAs (high-end laptops) switch to single stream mode (data redundancy) for protection against noise and interference. For that reason, even with 4 (or 8, which some call "*Wave 3*") antenna APs, the gains with respect to SU-MIMO (SDM mode) are modest. In SU-MIMO, the transmission to each STA is not simultaneous, but it exploits spatial multiplexing to deliver higher data rates. With 4 antennas, grouping is limited to 2 simultaneous users (and 4 with 8 antennas).

Finally, the most important problem about MU-MIMO is the fact that **channel probing protocol is expensive** in terms of airtime. Feedback matrices from the

---

[26] *Gesbert, David, et al. "Shifting the MIMO paradigm." IEEE signal processing magazine 24.5 (2007): 36-46.*

STAs to the AP are transmitted in series, resulting in high overhead. It has been said that, in changing environments, up to 85% of the transmissions may correspond to the channel probing protocol (wasting a lot of radio resources). Nevertheless, in the uplink, a *Trigger Frame* is needed to orchestrate the simultaneous transmission from STAs. Figure 28 shows the idea behind uplink MU-MIMO.

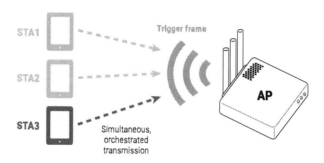

Figure 28. Uplink MU-MIMO. The Trigger Frame orchestrates the simultaneous transmission of all the devices simultaneously.

Uplink MU-MIMO was initially considered in 802.11ac, but it was not included due to implementation issues. Considering that the set of STAs in an MU group must jointly prepare their transmitters to send concurrent data to the AP, there are 3 main physical challenges for a successful uplink MU-MIMO transmission:

- Time synchronization of STAs.

- Frequency alignment of all the transmitter chains.

- Power normalization between all clients in an MU group.

The same happens for uplink OFDMA. A new protocol has been defined in 802.11ax to address these important issues, with novel frames (trigger) and frame grouping (multi-station ACKs) in order to reduce overhead and increase response time. In fact, **uplink and downlink multiuser transmissions may be OFDMA and/or MU-MIMO** and the Trigger Frame operates for all the options. Chapter 5 deepens in this feature.

Finally, groups have also been expanded (up to 8 users) for both downlink and uplink. As a result, even with modest devices in single stream mode, MU-MIMO throughput shows an improvement of 2x-3x over SU operation. Downlink and uplink MU-MIMO will be explained later in this book from the PHY (Chapter 4) and MAC (Chapter 5) point of views.

3.5 – Spatial frequency reuse

Another important problem in modern wireless communications is the scarce number of available channels. With classic frequency reuse, the system architect puts BSS (cells) with the **same channels as far as possible, in order to avoid intra and inter-channel interference**. To increase capacity in dense deployments, the frequency reuse between BSS needs to be increased. *BSS coloring* was a mechanism introduced in 802.11ah[27] (a standard that operates in the sub-GHz bands), and it will be extended to 802.11ax.

Until now, the *CCA (Clear Channel Assessment)* threshold establishes the minimum received signal power to keep quiet on the shared channel. Basically, if a WiFi transceiver detects a WIFi signal preamble with power over -82 dBm (or a non-WiFi signal with power over -62 dBm), it remains quiet waiting for another opportunity to transmit. However, **with modern receiver chains, a received power above this threshold may be allowed** without significant performance impact (this is known as *capture effect*). With this in mind, the idea behind BSS coloring is to establish a variable threshold that allows interfering signals if they come from BSS with a different *color*. This expands the frequency reuse of the entire network.

---

[27] *Toni Adame, Albert Bel, Boris Bellalta, Jaume Barcelo, and Miquel Oliver. IEEE 802.11ah: the*

WiFi approach for M2M communications. IEEE Wireless Communications, 21(6):144–152, 2014.

The <u>color</u> is a value which identifies groups of BSS which should not interfere. Thus, in 802.11ax a new channel access behavior will be assigned based on the color detected[28]. The new standard differentiates between intra-BSS frames and OBSS (Overlapping BSS) frames, with the use of *BSS color bits*. It applies a less sensitive CCA threshold to OBSS frames, so **their weak interference is (in some way) allowed**. Figure 29 illustrates the difference between classic BSS frequency reuse and modern BSS coloring.

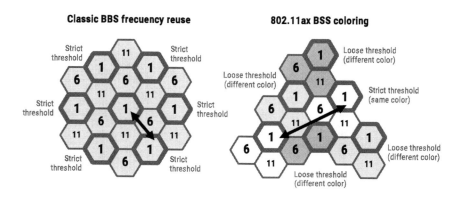

Figure 29. Classic BSS frequency reuse (left) and BSS coloring (right). Channel 1 has different thresholds for interference, depending on the BSS colors.

In 802.11ax, this threshold is dynamically adjusted. The impact on SNIR and MCS values is obvious, but usually, it is worth. So, a loose threshold **allows more simultaneous transmissions** (and wider channels) although potentially raises the interference.

---

[28] Villegas, Eduard Garcia, Rafael Vidal Ferré, and Josep Paradells. "Frequency assignments in IEEE 802.11 WLANs with efficient spectrum sharing." *Wireless Communications and Mobile Computing* 9.8 (2009): 1125-1140.

3.6 – High-order modulations

802.11ax allows the use of a 1024-QAM modulation. This means that one symbol carries 10 bits (instead of 8 bits/symbol with the 256-QAM). **Up to 25% more information in the same waveform.** Figure 30 compares both modulations for fixed signal, noise, and interference powers.

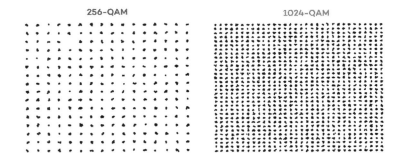

Figure 30. High-order modulations: 8 bits per symbol (left) vs 10 bits per symbol (right).

However, to maintain the same average/peak power (determined by the regulatory domain), the symbols must be nearer in the signal space (see Section 2.4.3). For a fixed SNIR, it is easier to distinguish between adjacent symbols in the 256-QAM than in the 1024-QAM. New MCS values (MCS10 and MCS11) appear to indicate this new modulation. Thus, the **SNIR shall be very high to trigger this new modulation,** so usually, it is reserved for STAs near to the AP and low-interference environments. Together with the new symbol duration, this may imply an improvement of about 40% in maximum data rates, with the same radio resources.

3.7 – Ready for outdoor environments

One of the goals of 802.11ax is to improve the outdoor performance, as well as providing long-range features. This will allow the expansion of WiFi for gardens, cities, stadiums, resorts, beaches, harbors, airports, and even highways. The guard interval of 802.11a/n/ac is 0.8 µsec long. However, in practice, this is not enough to deal with the multipath fading and inter-symbol interference caused by longer delay spreads.

The new standard introduces the **guard interval options of 1.6 and 3.2 μsec** (respectively, two and four times the duration of the original guard interval). In addition, a Doppler bit indicates *"Doppler mode of transmission"*, specifically designed for communication between high-speed vehicles. Table 4 summarizes the changes in terms of frequency-domain and time-domain features.

More features in 802.11ax related to the outdoor coverage and robustness are:

- New *"Extended range"* packet format with more robust preambles (STF and LTF are boosted by 3 dB, and L-SIG and HE-SIG-A fields are repeated twice) are introduced to expand the coverage and robustness of an outdoor hotspot. See Section 4.4 for more details.

- The *DCM (Dual Carrier Modulation)* replicates the same information on different subcarriers to implement frequency diversity, in order to gain narrowband interference protection (up to 3.5 dB). See Subsection 4.5.2 for more details.

- A narrow bandwidth transmission (8 MHz) is allowed for the Data Field in PHY frames, in order to reduce noise bandwidth and thus increase SNIR.

		802.11a/g/n/ac	802.11ax
Time domain	Guard interval	0.4 or 0.8 μsec	0.8, 1.6 or 3.2 μsec
	OFDM symbol	3.2 μsec	12.8 μsec
	Symbol time	0.4 μsec	12.8 μsec
	Efficiency	**80-89%**	**80-94%**
Frequency domain	FFT size in 20 MHz	64	256
	Subcarrier spacing	312.5 kHz	78.125 kHz
	Number of data subcarriers	48-52	234
	Efficiency	**75-81%**	**91%**

Table 4. Comparison of 802.11ax vs. 802.11a/g/n/ac PHY features related to long-range and outdoor environments.

3.8 – Focused on power efficiency

The main feature regarding power efficiency is the *TWT (Target Wake Time)* mechanism. It is inherited from the 802.11ah standard and consists of a negotiation between the STAs and its AP. **TWT allows the STA to sleep for periods of time, and to wake up in pre-scheduled times** in order to exchange information with its AP.

The AP schedules the target time. For IoT devices like sensors, which only transmits a little amount of data every few seconds, keep awake only to hear beacons is an unacceptable battery-drain behavior. With TWT, the sensor awakes in time and transmit its data, receives a response from the AP, and then goes back to sleep. The energy consumption is substantially reduced by this mechanism. The difference with 802.11ah TWT mechanism is the fact that 802.11ax TWT is fully compatible with triggered-based uplink transmissions (for OFDMA and MU-MIMO). In addition, there is also a Broadcast TWT mechanism for STAs that have no pre-negotiated an agreement with 802.11ax APs. TWT will be addressed in Subsection 6.3.2.

More power-saving features proposed in 802.11ax are:

- **Receive Operating Mode** indication: a Dynamic adaptation of the number of active receive chains and channel width (Transmit Operating Mode). It is a field in the MAC header in order to avoid the exchange of management frames.

- **Transmit Operating Mode** indication: a Dynamic adaptation of transmitting capabilities (spatial streams and bandwidth) of clients, depending on its energy policies. Dual to the Receive Operating Mode. It is also a field in the MAC header.

- **BSS Color field** in the preamble enables intra-PPDU power saving mechanisms, to distinguish intra BSS interference from external noise. BSS is described in Section 3.5 and Subsection 6.2.2.

3.9 – Conclusions

In this Chapter, we have presented the main issues of modern WiFi deployments. They are related to the overwhelming success of wireless communications, **so they shall be seen as opportunities instead of problems**. The ever-growing number of devices and things, as well as the increasing demand for higher data rates, imply a paradigm shift in terms of radio resource optimization and multiuser operation. And all of this without losing backward compatibility.

In this regard, we have seen that 802.11ax has been designed with these challenges in mind. Fortunately, the **new standard allows concurrent transmission and reception** by allowing multiple access in frequency (OFDMA) and space (MU-MIMO), both in the downlink and uplink directions. The next Chapters will delve into the insights of PHY and MAC layer features of 802.11ax.

4

802.11ax: PHY layer features

4.1 – Introduction

This Chapter delves into the main PHY changes introduced in 802.11ax. The new IEEE standard employs multiuser technologies (MIMO and OFDMA) in upload and download directions. For that purpose, **new scheduling methods are needed**. The signal processing part will be explained in the next sections, while the new protocols will be addressed in Chapter 5. In addition, we overview the proposal of the standard to achieve full-duplex (simultaneous transmission and reception) communications for the first time in WiFi history.

4.2 – Orthogonal Frequency Division Multiple Access (OFDMA)

4.2.1 – Resource Units

OFDMA is the multiuser variant of OFDM, where different sets of subcarriers are allocated to multiple users, allowing simultaneous access to radio resources. Until now, the whole channel was assigned to one specific user at a time. But with OFDMA, **portions of the spectrum are negotiated and assigned to different users.** This is something that has been done for mobile communications for the last years, due to the strong spectrum policies of 4G LTE technologies. Of course, this implies a pre-scheduling so every player knows where and when to transmit. For that purpose, new frames will appear.

The subcarriers of the BSS channel are divided into multiple groups[29], and each group is referred to as an *RU (Resource Unit)*. In this regard, **RUs are allocated to multiple STAs** depending on their channel conditions, service requirements, and hardware capabilities. Fortunately, the use of OFDMA reduces channel access overhead by amortizing those overheads and preambles across several STAs. It has another advantage: narrowband interference and deep fading impact are mitigated by avoiding these subcarriers.

4.2.2 – Uplink and downlink strategies

The ideas behind uplink and downlink OFDMA are similar:

- In the downlink, an AP raises the transmission power on some RUs (usually those corresponding to far or weak users), while reducing power for strong (or near) users in order to minimize interference. The aggregated transmitted power over all the band remains constant.

- In the uplink, OFDMA assigns RU to multiple users depending on their point of view over the radio channel, so there is an overall gain on robustness and aggregated throughput for the whole BSS. Normally, STAs transmit at a lower power than APs (because they need to save battery), so there is a power asymmetry that limits the BSS range. In this regard, OFDMA compensates such asymmetry, allocating smaller RUs for weak STAs allowing them to raise their SNIR (less in-band noise and interference).

4.2.3 – Subcarrier spacing and guard intervals

In 802.11ax, the subcarrier spacing has been reduced to gain granularity. It should minimize the guard interval overhead and provide better frequency-

---

[29] Deng, Der-Jiunn, et al. "IEEE 802.11 ax: Highly Efficient WLANs for Intelligent Information Infrastructure." IEEE Communications Magazine 55.12 (2017): 52-59.

selective gains. The selected value for **subcarrier spacing was 78.125 kHz** because it offers a good balance between granularity and spread protection. In addition, **guard interval durations of 0.8 μsec, 1.6 μsec, and 3.2 μsec** (2x and 4x the original length) protects transmissions against *ISI (Inter-Symbol Interference)* in a wide range of propagation delay values (indoor and outdoor environments). Table 5 shows the maximum number of RUs.

RU Type	20 MHz	40 MHz	80 MHz	160 MHz
26-tone RU	9	18	37	74
52-tone RU	4	8	16	32
106-tone RU	2	4	8	16
242-tone RU	1	2	4	8
484-tone RU	-	1	2	4
996-tone RU	-	-	1	2
1992-tone RU	-	-	-	1

Table 5. Maximum number of RUs for each channel bandwidth.

This implies **up to 9 simultaneous users in 20 MHz channels** and 18-74 for wider channels (which are not so usual in crowded environments). Figure 31 and Figure 32 show the RU distribution for 20 and 40 MHz bandwidths.

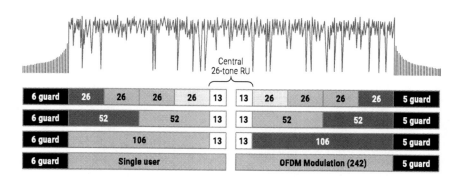

Figure 31. RU locations for 20 MHz channel bandwidth.

Note how the multiuser OFDMA degenerates in a single user OFDM modulation when we select wide RUs with all the tones in the bandwidth (242 for 20 MHz channels, and 484 for 40 MHz channels).

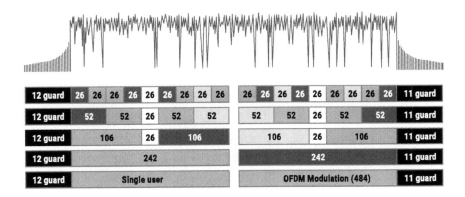

Figure 32. RU locations for 40 MHz channel bandwidth.

4.3 – MultiUser MIMO (MU-MIMO)

4.3.1 – Disaggregating the physical collision domains

The theoretical idea behind MU-MIMO has been well known for decades, but until recent advances in hardware, it has not been feasible. In order to transmit a symbol to a particular user (without interfering the others), we may **form a pattern of additive interference just at the point where that user is located, and another pattern of destructive interference where the others are.** This is known as beamforming[30] and, if we go no further, it has been available since the launch of the 802.11n standard (although rarely used).

---

[30] Litva, John, and Titus K. Lo. *Digital beamforming in wireless communications.* Artech House, Inc., 1996.

However, if we do that **for several users, adding the spatial interference patterns, we may send their data stream simultaneously.** As said, each user will only get the desired symbol, since the other symbols are attenuated by a pattern of destructive interference at that particular point in space. This is MU-MIMO and is available in the downlink thanks to 802.11ac Wave 2.

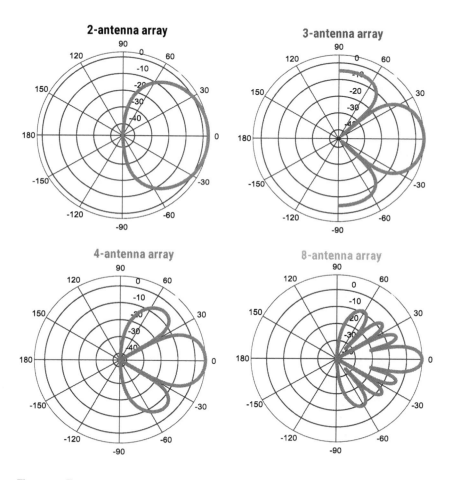

Figure 33. Energy patterns and beam widths for different antenna arrays (in dBi). The nearer to the external circle (0 dBi), the more power that the beam has in that direction.

For the first time in WiFi, the collision domain is divided. On an analogy, MU-MIMO converts WiFi shared access mechanisms from a hub (common collision domain) to a switch (disaggregated collision domains). This works, of course, if the terminals are far enough away from each other. The more antennas the transmitter has, the narrower than the beam can be. Figure 33 shows this feature of antenna arrays, and Figure 34 shows how this can be used as an enabling technology for MU-MIMO.

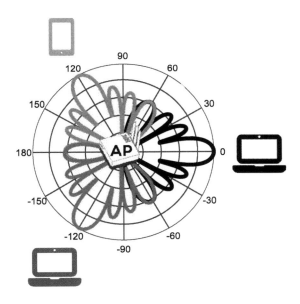

Figure 34. Downlink MU-MIMO by steering the beams of different data streams to different STAs.

4.3.2 – Signal processing and feedback

As stated in Figure 24, when transmit and receive antenna arrays are spaced apart far enough, **the multipath fading a signal suffers differs from one transmit-receive antenna pair to another.** The difference in channel quality between pairs of antennas is exploited either:

- To improve the robustness of the link (STBC or beamforming).

- To simultaneously transmit independent data streams from different transmit antennas to different transmit antennas (SDM).

However, the advantages of MIMO comes at the cost of requiring more complex signal processing and *CSI (Channel State Information)* gathering and feedback at the transmitter and/or receiver sides. In this regard, the overall process to start an MU-MIMO transmission is:

- In the <u>downlink</u>, also known as MIMO-BC (BroadCast):

  1. The AP probes the link between its antennas and the STAs.

  2. The STAs computes the CSI matrix and feedbacks it to the AP.

  3. The AP pre-codes its symbols by using the CSI matrix, in order to steer the data to each STA simultaneously.

- In the <u>uplink</u>, also known as MIMO-MAC (Multiple Access Channel):

  1. The AP probes the link between its antennas and the STAs.

  2. The AP announces to specific STAs that an uplink MU-MIMO transmission will be triggered soon.

  3. The AP sends a Trigger Frame to those specific STAs, which answers with an MU-MIMO data frame.

MIMO-MAC (uplink) requires CSI only at the receiver AP, which costs less in signaling overhead, as compared to MIMO-BC (downlink).

4.3.3 – More scheduling dimensions

As stated before, combining MU-MIMO and OFDMA multiple access frameworks, frequency, and spatial scheduling is available to the users. Uplink MU-MIMO improves the aggregated throughput of an IEEE 802.11ax network by **parallelization of multiple transmissions from STAs to the AP.** Specifically, it will be very useful for:

- Reducing the collision probability in the case of a large number of STAs.

- Long packet transmissions from multiple STAs.

Compared to OFDMA, MU-MIMO is **more suitable for STAs that are close to an AP with good receiving SNIR value and channel condition** (less noise and interference). However, both frameworks add system complexity in terms of the time, frequency, and power synchronization. They also introduce new probing and triggering protocols, so both sides of the link must support 802.11ax. The corresponding new frames will be presented in Chapter 5.

In an MU-MIMO resource unit, there is support for up to eight users with up to four space-time streams per user. However, the maximum number of space-time streams should not exceed eight, which means that at most two users will be using all of their streams. Table 6 summarizes the MU-MIMO configurations with a different number of antennas, spatial streams, and STAs.

# antennas In the AP	# STAs	# antennas per STA	Spatial streams	Beamforming	MU-MIMO
1	1	1	1	No	No
2	1	1	1	Yes	No
4	2	1	1	Yes	Yes, 2 sets of 2 antennas each
4	4	1	1	Yes, with 1 spatial stream	Yes, limited to 2 STAs, with 2 sets of 2 antennas each
8	2	4	4	Yes	Yes
8	4	2	2	Yes, with 1 spatial stream	Yes, 4 sets of 2 antennas each
8	8	1	1	Yes	Yes

Table 6. MU-MIMO configurations for different antennas and STAs.

4.4 – New PPDU formats

In Section 2.4.1 we introduced the PPDU format, which showed the structure of the PHY frames for the different versions of the 802.11 standard (non-HT, HT, and VHT). With 802.11ax, **five new PPDU formats are defined to support different modes of operation.** The preamble in all these new PPDU structures contains a *"legacy preamble portion"* to support backward compatibility (coexistence with legacy STAs). The reader may identify that the same decision was taken with regard to VT and VTH PPDU formats. This is followed by an *"HE-preamble portion"* (High Efficiency) to support 802.11ax enhanced features like OFDMA or MU-MIMO. Figure 35 shows these five new frame formats:

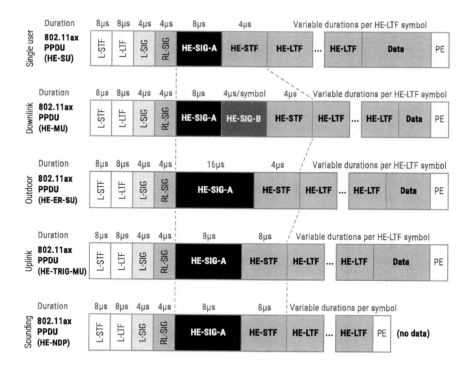

Figure 35. HE PPDU formats. RL-SIG stands for Repeated Legacy Signal Field (for redundancy), and PE stands for Packet Extension Field.

The following list details each format:

- **HE-SU PPDU**: it is used for single-user transmission only (to a single STA or to the AP), **like in non-HE** operation.

- **HE-ER-SU PPDU**: it is used for single-user, extended range transmissions (to a single STA or to the AP). Unlike other formats, this one contains an HE-SIG-A field that has a **repetition of each symbol (redundancy) and a power-boosted preamble** (L-STF and L-LTF) for reliable performance with longer coverage. It is limited to a single spatial stream and recommended for outdoors. Figure 36 shows the difference in signal strength of the HE-SU and HE-ER-SU preambles (L-STF and L-LTF).

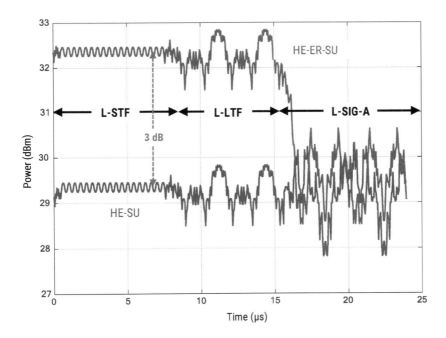

Figure 36. SU and ER-SU waveforms. The latter has stronger (3 dBm) training fields.

- **HE-MU PPDU**: it is used for multiuser transmissions (OFDMA or MU-MIMO) to one or more STAs. This requires the HE-SIG-B field assign one or more STAs in a PPDU. **It is mainly used for the downlink**, however, an STA may also transmit an HE-MU PPDU to the AP that supports its reception.

- **HE-TRIG PPDU**: it is used for triggered multiuser transmissions (OFDMA or MU-MIMO) from one or more STAs. **It is mainly used for the uplink**, as a response to the Trigger Frame, and shows a similar structure to the HE-SU PPDU, except in the fact that it uses a longer HE-STF field in the HE preamble. The information required for the uplink multiuser transmission from one or more STAs is not carried by the HE-SIG-B field, but by the Trigger Frame (see Subsection 5.3.3) that initiates this transmission.

- **HE-NDP PPDU**: a Null Data Packet is sent for the beamforming mechanism and **channel probing** (as in 802.11ac).

4.5 – Other PHY improvements

4.5.1 – In-Band Full-Duplex communications

IBFD (In-Band Full-Duplex) communications[31] is a technology that allows a transceiver to **simultaneously transmit and receive signals on the same channel**. Recently proposed, the main challenge to implement IBFD is how to perform *SIC (Self-Interference Cancellation)* from the transceiver own transmitted signal. This transmitted signal has been generated locally, so it has much higher power (various dozens of dBs) as compared to the desired received signal. Figure 37 illustrates the concept of SIC for IBFD communications.

---

[31] *Sabharwal, Ashutosh, et al. "In-band full-duplex wireless: Challenges and opportunities." IEEE Journal on selected areas in communications 32.9 (2014): 1637-1652.*

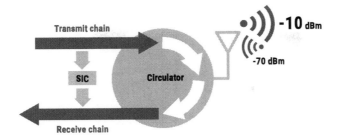

Figure 37. Self-Interference Cancellation in a wireless transceiver. The received signal is way weaker than the transmitted signal. Fortunately, the transceiver can subtract its contribution either in analog or digital way.

This can be achieved by using digital, analog, or mixed-signal processing. It depends on the hardware capabilities and performance and will increase the cost of devices. However, **this features will not be mandatory** in the release of the 802.11ax standard, so it every vendor may decide whether to include it or not. Fortunately, it will be backward compatible with previous 802.11 norms without additional changes.

4.5.2 – Modulation and Coding Scheme

As introduced in Section 3.6, the new 1024-QAM modulation achieves a 25% improvement on raw throughput for strong STAs, by bearing more bits per symbol than lower-order modulations. In this regard, 802.11ax **introduces MCS 10 and MCS 11 to enhance spectral efficiency in high SNIR environments**. They are optional for single-user and multiuser transmissions, but they are limited to RUs of 242 or more subcarriers. So, for 20 MHz channels, it collides with the OFDMA gain. Nevertheless, these MCS will only activate in high-end devices situated very near the APs.

Furthermore, in low SNIR environments, 802.11ax allows DCM (introduced in Section 3.7) to enhance robustness and reliability of transmissions in the presence of narrowband interference. DCM replicates **the same information in a pair subcarriers**, implementing *frequency diversity*. It is limited to MCS 0 to MCS 4 with one or two spatial streams. It is designed to trade throughput for the range.

Finally, 802.11ax forces the use of **LDPC codes** for several types of STAs:

- STAs supporting HE-SU PPDUs with bandwidths above 20 MHz.

- STAs supporting more than 4 spatial streams.

- STAs supporting 1024-QAM modulation. The coding rates are 3/4 for MCS 10, and 5/6 for MCS 11.

4.5.3 – 20 MHz-only operation

802.11ax takes into account the existence of low-power, low-throughput devices. In this regard, hardware manufacturers have now the permission to build specific 20 MHz-only clients, which is interesting for IoT devices (wearables, sensors, and actuators). Such devices do not need high data rates. Basically, **these devices only communicate on the primary channel**, while normal devices may spread to the whole 40, 80, or 160 MHz bandwidth. This promotes a cleaner spectrum with less interference for IoT. Figure 38 illustrates this mode of operation.

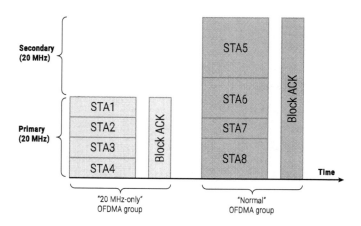

Figure 38. Operation with 20 MHz-only devices.

4.5.4 – Preamble Puncturing

In real-world deployments, **80 and 160 MHz bandwidth channels are difficult to implement** due to physical restrictions, high density of users, and interference. Specifically:

- The band of 5 GHz is unlicensed, but not contiguous. There is a set of channels known as *DFS (Dynamic Frequency Selection)* that cannot be used if there are radars working around. In practice, there are only five non-overlapping 80 MHz channels and one non-overlapping 160 MHz channels.

- The electromagnetic radiation regulations establish a limit for the maximum transmitted power. As the range of a BSS is extended (e.g., in outdoor WiFi networks), more interference is created and the set of available channels is further reduced.

- Most of legacy APs operates with 20 MHz channels and are deployed along all the band, decreasing the white spaces to expand your bandwidth.

To overcome these issues, 802.11ax introduces *preamble puncturing*, allowing an AP to transmit an HE-MU PPDU (downlink) in punctured 80 or 160 MHz channels **when part of their 20 MHz sub-channels are busy**. These occupied 20 MHz sub-channels must not be the primary channel.

This feature is optional for both AP and STAs, but it must be announced in the capabilities information element. After negotiating this mode of operation, **the preamble part will be punctured**: it will be not be transmitted in the busy 20 MHz sub-channels. This will enhance the channel utilization for the high-interference scenarios where wider bandwidths may not be fully available all the time.

The logic behind channel puncturing is the following:

- **For 80 MHz transmissions**, only one of the 20 MHz sub-channels other than the primary 20 MHz channel will be punctured.

- **For 160 MHz transmissions**, one of these options may happen:

  o The secondary 20 MHz sub-channel will be punctured in the primary 80 MHz channel.

  o The two 20 MHz sub-channels corresponding to the primary 40 MHz channel will not be punctured. At least one of the other 20 MHz sub-channels corresponding to the 160 (or 80+80) MHz channel will be punctured.

In any case, we will have to see how preamble puncturing behaves in real-world WiFi deployments, in terms of reliability and throughput.

4.6 – Conclusions

In this Chapter, we have explained the PHY features that 802.11ax brings to WiFi networks. There are some interesting changes like the use of high-order modulations or new PPDU formats. But the main contribution comes from OFDMA and MU-MIMO. The combination of these multiuser access methods **enables, for the first time in WiFi, two additional dimensions for scheduling users: frequency and space.**

Until now, airtime was given to users by a random, simple, and inefficient access method. However, the introduction of OFDMA and MU-MIMO (both in uplink and downlink) implies the interchange of new frames in order to orchestrate the resources assigned to each STA. These frames appear at the MAC level, and they will be explained in the next Chapter, as well as other Layer 2 improvements.

5

802.11ax: MAC layer features

5.1 - Introduction

This Chapter is built upon the previous one. In fact, the MAC layer is located just above the PHY layer in the *OSI (Open System Interconnection)* stack. The MAC layer is the lower part of the 2-layer, and the 802.11 standard defines its operation together with the 1-layer (PHY). In order to implement all the new functionalities presented in Chapter 4 for 802.11ax, new frames are defined.

5.2 – Overview of MAC features

OFDMA and MU-MIMO create a fundamental challenge in 802.11ax MAC design due to the multiple users sharing a frequency channel, resulting in the introduction of novel carrier sense methods. In addition, new protocols have been designed in both upload and download, in order to **orchestrate the concurrent transmissions**. The new MAC methods must be compatible with original CSMA-CA, and it is based on the concept of RU (See Subsection 3.3).

5.2.1 –Overview of Downlink Multiuser Modes

The HE-MU format defined in Subsection 4.4 can be used in the downlink for an OFDMA transmission, an MU-MIMO transmission, or a combination of the two. This flexibility allows an HE-MU packet to transmit from the AP to:

- A **single STA** over the whole channel (as in previous 802.11 amendments). Figure 39 shows this simple RU allocation.

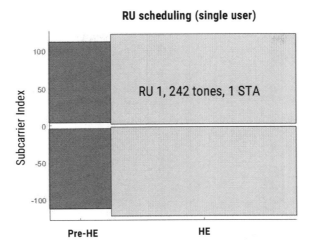

Figure 39. RU allocation when the AP transmits to single STA using all the channel.

- Multiple STAs over different zones of the channel (**OFDMA**). The channel bandwidth is divided into RUs (a group of subcarriers), which are defined by the size and the index. RUs were defined in Section 4.2.1. Figure 40 shows this RU allocation with 4 target STAs.

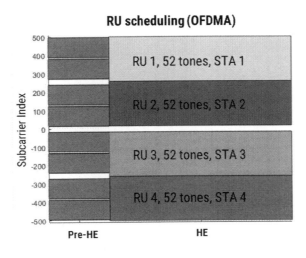

Figure 40. RU allocation when the AP transmits to 4 STAs by using OFDMA.

- Multiple STAs over the whole channel (**MU-MIMO**). For example, an AP transmitting simultaneously to four STAs, like in 802.11ax Wave 2. Figure 41 shows this RU allocation with only one RU and 2 STAs.

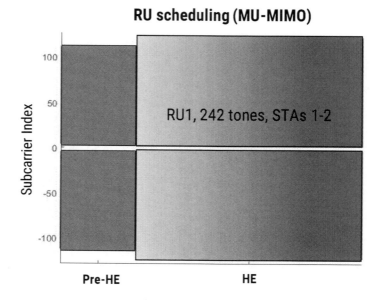

Figure 41. RU allocation when the AP transmits to 2 STAs by using MU-MIMO (like in 802.11ac Wave 2).

- Multiple STAs over the same part of the channel (**Mixed MU-MIMO and OFDMA**). For example, two users share a specific RU (MU-MIMO), and the other two users share another RU (MU-MIMO). Of course, together they form an OFDMA scheme. Figure 42 shows the RU allocation for this configuration. Note that the bandwidth assigned to each STA is higher than in Figure 40.

The next section explains the same examples of the use of OFDMA and MU-MIMO but in the uplink direction: from the STAs to the APs. The uplink multiuser transmission is triggered and coordinated by the AP, while there are modes of downlink multiuser transmission that does not need a trigger. In order to illustrate the duality between RU allocation and spectrum, for the downlink **we have selected the same four examples** than in the previous list. In this way, the reader may identify the resource allocation structure in Subsection 5.2.1 with the spectral shape of the Subsection 5.2.2.

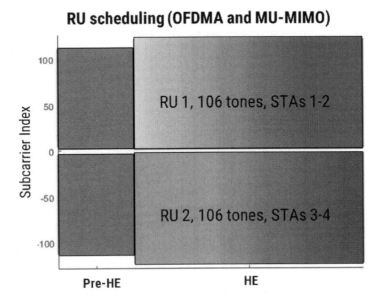

Figure 42. RU allocation when the AP transmits to 4 STAs by using OFDMA and MU–MIMO together.

5.2.2 – Overview of Uplink Multiuser Modes

In this case, the HE-TRIG-MU format is used as a response to the stimulation received by a Trigger Frame from the AP (which will be addressed in the Subsection 5.3.3). With the same options described in Subsection 5.2.1, the uplink transmission modes are:

- From a **single STA** (as in previous 802.11 amendments). Figure 43 shows the spectrum of this configuration, which has the same RU allocation than in Figure 39.

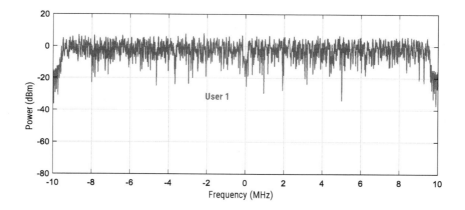

Figure 43. The spectrum of a single STA using all the channel.

- Multiple STAs over different zones of the channel (**OFDMA**). Figure 44 shows the spectrum with this configuration with 4 users, which has the same RU allocation than in Figure 40.

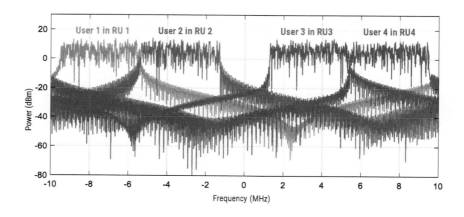

Figure 44. Spectrum of an OFDMA scheme with 4 STAs and 4 RUs.

- Multiple STAs over the whole channel (**MU-MIMO**). This option was not available in 802.11ac Wave 2 for the uplink. Figure 45 shows the spectrum with this configuration and 2 users, which has the same RU allocation than in Figure 41.

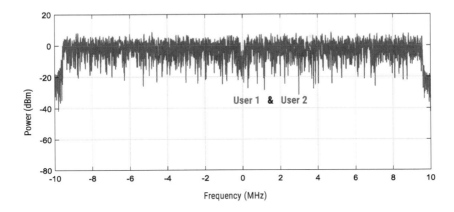

Figure 45. The spectrum of 2 STAs sharing the channel thanks to MU-MIMO.

- Multiple STAs over the same part of the channel (**Mixed MU-MIMO and OFDMA**). Figure 46 shows this configuration for 4 users and 2 RUs, which has the same RU allocation than in Figure 42.

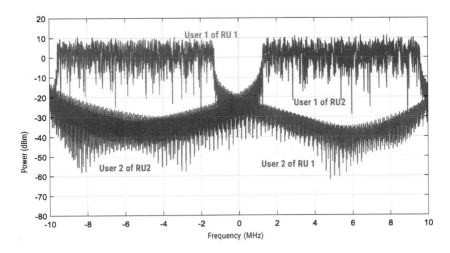

Figure 46. The spectrum of 4 users sharing 2 RUs (OFDMA and MU-MIMO together).

5.3 – Preparing the Data Transmission

Before describing how the data transmission is done in 802.11ax simultaneously for multiple users, this Section presents some mechanisms that help to configure and orchestrate a successful transmission with both OFDMA and MU-MIMO paradigms.

Both downlink and uplink multiuser transmissions are scheduled by the AP through *Trigger Frames*, which are in charge of reporting instructions to the involved STAs. Trigger Frames will be deeply described and analyzed in Section 5.3.3.

5.3.1 – Winning the Multiuser Channel

In CDMA-CA, the channel is won by sending an RTS frame and receiving back a CTS frame. **This frame is not always necessary**, although it helps to reduce collisions and retransmissions in the BSS. For the multiuser transmission, an MU-RTS frame was defined in 802.11ac Wave 2, in order to guarantee the same behavior from multiple devices.

The MU-RTS frame[32] requests simultaneous CTS responses from multiple STAs to win the channel in order to start a protected downlink multiuser transmission. **It is similar to the Trigger Frame**, so the MU-RTS frame provides:

- Time synchronization among target STAs.

- Frequency offset correction based on the common AP reference.

- Signals to identify the target STAs.

---

[32] *Liao, Ruizhi. "Mac design and analysis for mu-mimo and full-duplex enabled wireless networks". Diss. Universitat Pompeu Fabra, 2014.*

As a response, the CTS will be transmitted simultaneously (multiplexed with OFDMA or other orthogonal mechanism) from multiple STAs to reduce overhead. It must use exactly the same transmission parameters. The procedure is illustrated in Figure 47.

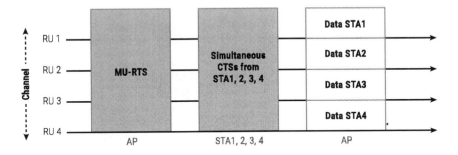

Figure 47. MU-RTS and simultaneous CTS frames.

5.3.2 – Probing the Channel for Beamforming

The following transmission method has been present in 802.11ac since 2016 (downlink only) when the Wave 2 devices arrived at the market. We outline in this Section the main steps, as it is also present in 802.11ax HE-MU mode. **An MU-MIMO transmission starts by sounding the radio channel between transmitter and receiver.** With the acquired CSI, the transmitter or receiver can adapt its PHY parameters in order to provide an orthogonal framework for the simultaneous transmission or reception of data frames.

To probe the medium, a protocol is proposed that characterizes the radio channel and acts accordingly, giving instructions to each transmitter chain. **Transmit beamforming for MU-MIMO is based on the knowledge of the CSI between transmitter and receiver at the AP.** In addition, OFDMA RU allocation may be based on the channel conditions between AP and STAs. By channel probing, feedback of the subcarrier CSI is provided by each STA. Figure 48 shows the frames interchanged to probe the channel before a downlink MU-MIMO transmission.

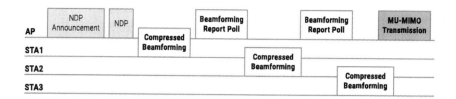

Figure 48. Channel probing, feedback, and adaptive beamforming to steer information for each user.

The process consists of a series of frames requesting and gathering information, and it is similar for both OFDMA and MU-MIMO paradigms:

1. The AP sends an **NDP (Null Data Packet) Announcement** frame, which notifies that the channel probing procedure will start soon. Figure 49 shows the structure of the NDP Announcement frame. The Receiver address is the broadcast address. The most important fields are the STA Info fields (one for each target STA), which include the Association ID (AID12) of the client device, the feedback type, and the number of spatial streams.

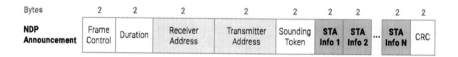

Figure 49. Structure of the NDP Announcement frame (from AP to the STAs).

2. The AP sends an **NDP** frame, which does not contain information from higher levels to the different users. These frames intend to probe the radio channel. The NDP frame is just a single user PPDU without frame body.

3. The target STA decode the frame header and, by using advanced signal processing techniques (SVD - Singular Value Decomposition), extract relevant information about how the channel behaves at that specific time.

4. By using the **Compressed Beamforming Report** frame, the first STA sends compressed information to the AP, with instructions about how to steer the beam (by electronic means) to create a constructive pattern in its direction. Figure 50 shows the structure of this frame. The most important fields are the MIMO Control field (that enables a beamformer to interpret the feedback matrix by describing its size, channel width, grouping, and flow control information), and the Compressed Beamforming Report itself (that codes the phase shifts required by each antenna of the array). The MU Exclusive Beamforming Report field is not present for a single-user transmission and carries SNR differences between subcarriers in order to update the steering matrix for multiple STAs. The variable size depends on the number of spatial streams and on the channel bandwidth.

Figure 50. Structure of the Compressed Beamforming Report frame (from the STAs to the AP).

5. The AP must send **Beamforming Report Poll** frames in order to gather the Compressed Beamforming frames from the other STAs. This frame is shown in Figure 51. This frame is quite simple and presents a bitmap where flags for polled STAs are activated, while the rest are deactivated. There is a type of Trigger Frame which is indeed a Beamforming Report Poll Frame.

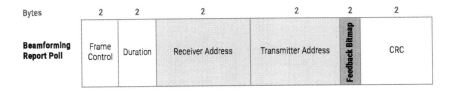

Figure 51. Structure of the Beamforming Report Poll frame (from AP to the STAs).

6. The AP receives this information matrix from all the users to be beamformed (the *beamformees*), and adapts its RF chains for that purpose: the transmitter multiplies each symbol assigned to a user by the calculated interference pattern.

Note that this complex process of adaptive probing has an important computational cost, but it is also expensive from the protocol and spectral efficiency point of view. Nevertheless, the gathering of CSI implies several advantages, and it will be present in the next-generation wireless communication standards for several years.

5.3.3 – The Trigger Frame

The AP may orchestrate and trigger the simultaneous transmission from all the STAs, as well as provide instructions to the STAs before transmitting a multiuser frame (mainly the allocated RUs). As the uplink procedure is slightly more complicated, this Section describes how to trigger a multiuser uplink transmission, regardless of the use of MU-MIMO, OFDMA, or a combination of both paradigms.

The basic idea consists of a *Trigger Frame* that starts a four-way handshaking procedure before multiuser transmissions. The Trigger Frame is a Control Frame, which is sent from the AP (who has won the channel) to multiple STAs. Its purpose is **to trigger the user devices** to transmit an HE-TRIG-MU PPDU in the uplink direction.

The uplink HE-TRIG-MU PPDU bears a Data Frame, transmitted as a response to the Trigger Frame was designed with a split structure:

- A pre-HE part, which is the same for all the HE-TRIG-MU frames.

- An HE part, where all HE LTF would be mutually orthogonal (in space or frequency) between any two of the triggered STAs. Each of the triggered STAs form its own HE LTF sequence based on the STA order.

An AP can distinguish each user by using this orthogonality when receiving HE-TRIG-MU PPDUs. Finally, **the AP estimates the uplink CSI between the AP and each STA**, in order to successfully decode each received frame. The benefits of this design are twofold:

- It can let an AP treat the received signal as being from one transmitter.

- It can protect the transmission by using the legacy preamble to report necessary information about the current PPDU to non-HE STAs.

The complete structure of the Trigger Frame is shown in Figure 52.

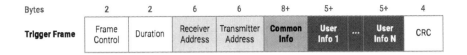

Figure 52. Trigger Frame structure.

The frame control has the same subfields than in other 802.11 frames. The duration and the addresses (RA is the broadcast address) are found before frame-specific subfields. In the end, there is an optional padding before the CRC. A Common Information field carries the shared information for all the triggered STAs. Its structure is shown in Figure 53 where the most important bits are shaded.

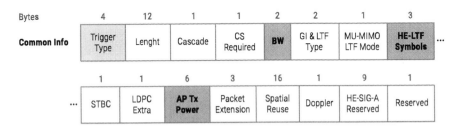

Figure 53. Common Information Field structure.

The subfields of the Common Information field are described below. Most of them set specific parameters on the triggered frame (the uplink Data Frame transmitted as a response to the trigger frame):

- The **Trigger Type** sets one of the following types: Basic, Beamforming Report Poll, MU-BAR (Block ACK Request), MU-RTS, and Buffer Status Report Poll, among others.

- The **Length** sets the value of the L-SIG length field of the HE trigger-based PPDU.

- The **Cascade** bit indicates a subsequent Trigger Frame following the current one.

- The **CS Required** bit is activated to indicate that the STAs must sense the channel when determining whether or not to respond.

- The **Bandwidth** bits set the bandwidth in the HE-SIG-A field of the triggered uplink frame.

- The **GI and LTF Type** bits set the Guard Interval and Long Training Field of the triggered frame.

- The **MU-MIMO LTF Mode** bit set the LTF Mode of the triggered frame (0 for single stream pilots, or 1 for masked LTF sequence of each spatial stream by a distinct orthogonal code).

- The **Number of HE-LTF Symbols** bits present in the triggered frame.

- The **STBC** bit enables the Space-Time Block Coding on the triggered frame.

- The **LDPC Extra Symbol** bit enables the Low-Density Parity-Check code extra symbol segment.

- The **AP Tx Power** bits set the combined average power per 20 MHz bandwidth of all transmit antennas.

- The **Packet Extension** bits set the packet extension duration of the triggered frame.

- The **Spatial Reuse** bits set the value for the spatial reuse field in the HE-SIG-A field of the triggered frame.

- The **Doppler** bit indicates a high Doppler mode of transmission, specifically used for moving users and things.

- The **HE-SIG-A Reserved** bits set the value of the reserved bits in the HE-SIG-A field of the triggered frame.

In addition, the <u>User Info</u> field is present for each one of the target users. It consists of 5 fixed Bytes, with more optional bits, per user. It is shown in Figure 54.

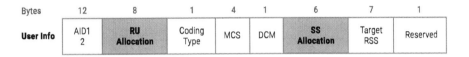

Bytes	12	8	1	4	1	6	7	1
User Info	AID1 2	**RU Allocation**	Coding Type	MCS	DCM	**SS Allocation**	Target RSS	Reserved

Figure 54. User Information Field structure.

The following subfields are always present in the User Info field:

- The **AID12** bits set the least significant 12 bits of the identification number of the STA for which the User Info field is intended. **An AP might assign the RUs for *random access*[33] by setting this value to 0 in this subfield.** When an STA receives a Trigger Frame with random access, it should reduce its OFDMA back-off counter. Then, the STA could randomly select one of the available RUs to perform uplink OFDMA. This is especially recommended for very short burst transmissions, for unassociated STAs,

---

[33] *Afaqui, M. Shahwaiz, Eduard Garcia-Villegas, and Elena Lopez-Aguilera. "IEEE 802.11 ax: Challenges and requirements for future high efficiency WiFi." IEEE wireless communications 24.3 (2017): 130-137.*

and also in the case that the AP aims to fully use the unallocated RUs in uplink OFDMA.

- The **RU Allocation** bits set the RU allocation (time slots and frequency subcarriers) used by the triggered frame.

- The **Coding Type** bit sets the channel coding (block or LDPC) of the triggered frame.

- The **MCS** bits set the Modulation and Coding Scheme of the triggered frame.

- The **DCM** bit sets the dual carrier modulation of the triggered frame.

- The **SS Allocation** bits set the spatial streams of the triggered frame.

- The **Target RSSI** bits set the target received signal power of the triggered frame (Values 0 to 90 are mapped to -100 to -20 dBm, with 1 dB resolution). Value 127 indicates the STA to transmit at the maximum allowed power for the MCS set.

Finally, additional bits are allowed, like the **Feedback Segment Retransmission Bitmap,** which sets the requested feedback segments of an HE Compressed Beamforming report (Only for Beamforming Report Poll Trigger Frame).

5.4 – The Multiuser Data Transmission

Once the transmission has been configured and prepared, this Section explains how the user data is transmitted in both downlink and uplink directions.

5.4.1 – Downlink OFDMA and MU-MIMO

In the downlink, after the channel is known at the transmitter, the AP steers simultaneous frames to each STA (by using OFDMA, MU-MIMO, or a mixed mode). Later, it triggers the ACK from the receivers by sending specific **Block ACK Request (BAR)**, which are responded by the STAs with a Block ACK if the frame was successfully received. MU-BAR is also a type of Trigger Frame. The procedure is illustrated in Figure 55, and it has the same structure for both downlink OFDMA and MU-MIMO.

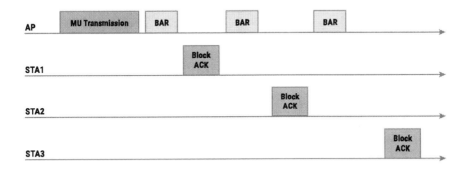

Figure 55. Downlink MU transmission and acknowledgment.

Note that **both sides of the communication (the AP and the STAs) need to be aware of this protocol** in order to successfully perform the multiuser operations like channel probing, feedback, and steering. As the new standards penetrate the market, we will see more devices supporting these advanced features in subsequent Waves.

5.4.2 – Uplink OFDMA and MU-MIMO

Before triggering an uplink OFDMA, MU-MIMO, or mixed multiuser transmission, **the AP collects the requirements of each STA**. Of course, all the triggered frames shall have the same length. In order to determine if the STAs are ready to send frames up, two complementary mechanisms[34] are

---

[34] *Bellalta, Boris, and Katarzyna Kosek-Szott. "AP-initiated Multi-User Transmissions in IEEE 802.11 ax WLANs." arXiv preprint arXiv:1702.05397 (2017).*

introduced to allow STAs to send *BSRs (Buffer State Reports):* solicited and unsolicited, depending on the acceptable cost overhead. The *BSR Poll frame* is a type of Trigger Frame that the AP sends to interrogate STAs.

After receiving the Trigger Frame, each of the triggered STAs complete the following procedure in a fixed amount of time:

1. Synchronization with the Trigger Frame, including time (clocks) and frequency (carriers) errors.

2. If the CS Required bit is active, check if the channel is free (energy detection threshold).

3. Adapt the transmitted power based on the parameters in the Trigger Frame (AP Tx Power and Target RSSI).

4. Prepare PPDU as the PHY parameters indicated in the Trigger Frame.

Then, the signals are transmitted simultaneously by the STAs, and they reach the antennas of the AP. Fortunately, they do not collide as they transmit only on the resource units established in the RU Allocation subfield. **If MU–MIMO is also used**, the receiver (AP) antenna arrays are electronically oriented to maximize the different incoming beams by using the steering matrix. Figure 56 shows this behavior.

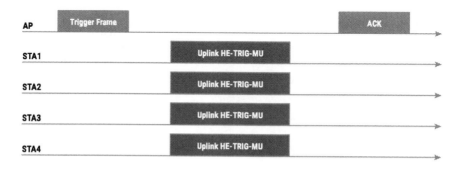

Figure 56. Uplink trigger-based MU transmission and acknowledgment.

Finally, the AP has to send an acknowledgment in response to the uplink multiuser transmission. It has to decide whether to send an individual ACK to each STA (by using downlink OFDMA) or one broadcast frame (Block ACK), incorporating the response to all the STAs in a single transmission.

5.5 – In-Band Full-Duplex MAC

As stated in Section 4.5.1, IBFD communications break the fundamental hypothesis of 802.11 MAC. That is, that a device cannot transmit and receive simultaneously. Thus, this new full-duplex mode may **improve substantially the overall throughput of the BSS**. However, before including IBFD in 802.11ax scenarios, some issues need to be considered:

- What happens when the transmitted and received frames have different lengths?

- How does this affect the standard RTS/CTS scheme for sensing if the channel is free?

- How and when do the devices interchange ACKs after finishing an IBFD transmission?

- A third device not involved in the IBFD communication will receive two different frames, as there are at least two simultaneous transmitters in the same BSS. Thus, airtime fairness is reduced.

- What happens to multiuser transmissions? If an AP wants to communicate with two STAs by using IBFD, each STA may interfere with the other.

For these and other issues[35], **it is possible that IBFD communications will not be present in the first Wave of 802.11ax devices.** Hopefully, in the final release, IBFD shall remain optional for device manufacturers. It is possible, then, that high-end vendors will implement this feature, although it will not be soon.

5.6 – Conclusions

In this Chapter, we have presented 802.11ax MAC strategies and features. The major advance at layer 2 is the presence of trigger-based multiuser transmission mechanisms, like OFDMA and MU-MIMO. While downlink multiuser transmission has been supported by 802.11ac Wave 2 through MU-MIMO, **the uplink multiuser transmission has not been considered by any previous IEEE 802.11 amendment**. It remains a challenging effort, mainly due to the requirement to synchronize transmissions from multiple STAs to the AP. We reviewed how to schedule RUs and the structure of the frames involved in the uplink and downlink multiuser methods.

---

[35] *Omar, Hassan Aboubakr, et al. "A survey on high efficiency wireless local area networks: Next generation WiFi." IEEE Communications Surveys & Tutorials 18.4 (2016): 2315-2344.*

6

Radio Resource Management

6.1 – Introduction

This Chapter summarizes the main results and efforts of 802.11ax in regard to radio resource management and optimization. Concretely, there are some **power and frequency-related enhancements** in order to make a better use of the limited resources in WiFi communications. Some of the most interesting methods were introduced in Sections 3.5 and 3.8.

6.2 – Spatial Frequency Reuse

In dense deployment scenarios, it is particularly important for HEWs to manage interference among neighboring BSSs. The following Subsections describe the proposed methods for efficient reuse of the available channels.

6.2.1 – Enhanced Clear Channel Assessment

As stated in Section 3.5, current CCA method described in 802.11 uses fixed thresholds to determine whether the channel is busy or idle. The device which detects the wireless channel as busy defers its channel access, in order to avoid interference. The CCA levels are set to low values for three reasons:

- To increase the communication range among STAs.

- To decrease the interference power level at the receiver STA.

- To minimize *hidden terminals*, thanks to the RTS/CTS mechanism.

However, this feature was designed for a world different than the world of today. In a dense deployment scenario like a convention center or an aircraft, **a low CCA threshold may cause an innocent transmission in a BSS to prevent many STAs in the surroundings from using the channel**, which unacceptably degrades the overall throughput.

The *exposed terminal* problem is more subtle than the hidden terminal problem. In this case, a terminal may be silenced (due to a bad selection of CCA thresholds) even if it could be transmitting with acceptable SNIR.

The basic idea to overcome problems related to a fixed CCA threshold is to use a *DSC (Dynamic Sensitivity Control)* scheme, in order to optimize the performance of the WiFi network by **fine-tuning of the CCA threshold**. This shall be done in a distributed, robust, and adaptive manner, for each node. Of course, the values must remain upper and lower bounded to avoid unstable behaviors.

For that reason, several CCA enhancement has been proposed for 802.11ax. It is possible that the final selected option will remain open for the device manufacturers, so the standard will not define a specific new CCA method. Each AP or STA manufacturer may use the method (or methods) that best suits their needs, as long as they satisfy the standard and backward compatibility concerns.

6.2.2 – BSS Color and multiple timers

BSS Color establishes different thresholds for signals coming from inside and outside the BSS. It has been successfully implemented in 802.11ah deployments, and now it is time to adopt it in 802.11ax networks. We addressed the basic mechanism in Section 3.5, which we recommend the reader to review.

With BSS Color, receivers apply a higher CCA threshold to OBSS frames, so **weak interference is not forbidden**. Upon receiving frames from an OBSS, a station can cancel the reception process (assuming that the channel is idle), thus increasing its transmission opportunities.

Regarding timers, it is well known that the WiFi devices, upon receiving the RTS/CTS frames, set a timer that blocks them from transmitting during a specific time. This timer is the *NAV (Network Allocation Vector)*. However, in 802.11ax, the use of **various different NAV timers per station has been proposed**. Each STA or AP maintains multiple NAVs, each one corresponding to a particular BSS head by the device (one of them for its own BSS).

The key idea is that a NAV corresponding to a particular BSS can be reset or increased only by reception of frames from that BSS. **If at least one NAV indicates that the medium is busy, then the medium is considered to be busy.** This is compatible with the use of dynamic CCA thresholds. Figure 57 illustrates an example where multiple NAVs avoids a collision from single NAV expiring. It can be seen that, without the existence of NAV2, the AP2 will transmit after NAV1 is reset (after the CF-end of BSS1). Thanks to multiple NAVs, the AP2 is inhibited from transmitting because of NAV2.

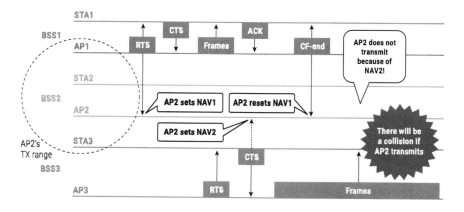

Figure 57. Multiple NAVs to avoid collisions between OBSS frames.

As the reader may note, **most of the interference-related enhancements are based on differentiating intra-BSS from OBSS frames** in terms of time, frequency, and signal strength. It is expected that 802.11ax real-world networks will have coordination mechanisms to activate and deactivate in real-time these features for different devices. Possibly, these mechanisms will be triggered by the detection of hidden and exposed nodes, high levels of interference, and other performance indicators.

6.3 – Power Efficiency

This Section describes the new features that 802.11ax engineers and scientists are considering for minimizing the power consumption, maximizing the battery life of handheld devices, as well as reducing interference and radiation.

6.3.1 – Transmit Power Control

802.11ax is contemplating the use of per-link *TPC (Transmit Power Control)*. It consists of **dynamically changing the power of transmitted frames for each specific AP-STA pair**. Until now, power control is the same for all the STAs within a BSS, and cannot be defined for specific STAs. A per-link TPC mechanism may reduce interference as well as increase spatial reuse.

The aim of TPC is to dynamically adjust the lowest possible power for stations with the weakest signal (usually the furthest), in order to reach an SNIR enough to decode the received frames at the desired MCS.

6.3.2 – Target Wake Time

As introduced in Section 3.8, power consumption is a major issue in modern WiFi networks[36]. In this regard, TWT is a mechanism that allows STAs to **sleep during a certain time in order to save battery**.

---

[36] *Nurchis, Maddalena, and Boris Bellalta. "Target Wake Time: Scheduled access in IEEE 802.11 ax WLANs." arXiv preprint arXiv:1804.07717 (2018).*

TWT has a significant role in mitigating the negative effects of channel contention, by providing a **simple but effective mechanism** to schedule transmissions in time. It is considered a major step towards a practical orchestrated, collision-free, deterministic access in future WiFi networks. If this is achieved, the service might be guaranteed for the first time in WiFi networks.

TWT, like other 802.11ax features, was introduced in 802.11ah amendment aiming to provide a low-consumption mode for stations with low traffic requirements, and periodic data transmissions. It fits perfectly with the IoT and Edge Computing paradigms, where a significant portion of network and processing tasks are done at the edge (where the data is generated), and thus **there is a need for orchestration and scheduling** in order to save battery and optimize resources.

Although it might be counterintuitive, reception chains consume battery as well as transmission chains do. By using TWT, each station negotiates awake periods with the APs to transmit and receive frames, while **saving energy the rest of the time** as the station remains in sleep mode. This has two main advantages for WiFi networks:

- Leads to a low energy consumption for the STAs, saving power and reducing electromagnetic radiation.

- Reduces the interference level significantly, pursuing a collision-free and deterministic operation.

Therefore, the use of TWT can result in significant throughput gains in high-density, high-demanding scenarios. However, a proper awake/sleep mechanism is needed in order to give to each traffic flow the required airtime and opportunities.

TWT offers two modes of operation:

1. **Individual TWT mode:** STAs negotiate a common wake scheduling with the AP. This allows them to **wake up only when required**, minimizing power consumption and interference. The parameters negotiated include:

   - The next time in microseconds at which the STA should wake up.

   - The time interval between subsequent TWT sessions for the STA.

   - The minimum time an STA shall stay awake to be able to receive frames.

   - The channel that STA can use temporarily as the primary one.

   - The mechanism employed to protect a TWT session from external RTS/CTS and other transmissions.

2. **Broadcast TWT mode:** allows an AP to set up a shared TWT session for a group of STAs, and **specify periodically the parameters by using Beacon frames**. In this scenario, the STAs are required to wake up to receive only the Beacons with instructions. It is interesting to note that STAs may request membership to existing TWT sessions, which are announced by the APs.

There are two **roles in TWT**: *Requesting* (initiates the setup of the session) and *Responding* (accepts or rejects the request). There are also different message types for negotiating parameters, as Suggest/Request/Demand on the requester side, and Accept/Alternate/Dictate/Reject on the responding side. When the final response is of type *Accept*, the session has been configured and established.

The STAs goes to the sleep state and wakes up at the time at which the next relevant Beacon is scheduled, and these Beacons carry the necessary information about the session, in order to follow the session timing and schedule.

TWT sessions may be triggered or not by the AP. In addition, the session may be specified as *Announced* (the STAs have to send PS-Poll messages to the AP to require buffered data) or *Unannounced* (allows the AP to deliver data without waiting for any previous frame STAs). Figure 58 illustrates the Triggered and Announced Broadcast TWT mechanism. The STAs send a *PS-Poll (Power Save Poll)* frame to query the status of the negotiated sleep period.

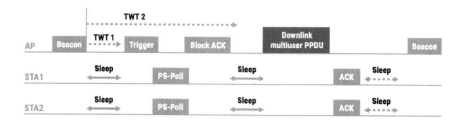

Figure 58. Target Wake Time mechanism.

6.4 – Conclusions

This Chapter has been dedicated to the description of various 802.11ax features that will enhance the spectrum and power usage of the devices. The aim of the presented mechanisms is to **reduce power consumption and interference, as well as pursuing a collision-free shared transmission medium.** Enhanced CCA mechanisms, together with BSS coloring and multiple NAVs can reduce substantially the problems of CSMA-CA without losing backward compatibility with old devices. In addition, TWT allows the orchestration of sleep and awake times in order to reduce energy consumption.

Some of these new features will be mandatory, and others will be optional. During the next months, hardware manufacturers and device vendors will decide which optional features include in their products. The releases will be probably in two or three different Waves.

Index of Figures

Printed in Poland
by Amazon Fulfillment
Poland Sp. z o.o., Wrocław